Design & Art Hotel

设计艺术酒店

第二版

李壮 主编

中国林业出版社
China Forestry Publishing House

图书在版编目（CIP）数据

设计艺术酒店 / 李壮主编 . -- 2 版 . -- 北京 : 中
国林业出版社 , 2017.9

ISBN 978-7-5038-9248-6

Ⅰ . ①设… Ⅱ . ①李… Ⅲ . ①饭店－室内装饰设计－
中国－图集 Ⅳ . ① TU247.4-64

中国版本图书馆 CIP 数据核字 (2017) 第 207228 号

中国林业出版社 · 建筑分社
策　　划：纪　亮
责任编辑：纪　亮　王思源
封面设计：吴　璠

出版：中国林业出版社（100009 北京西城区德内大街刘海胡同 7 号）
网站：lycb.forestry.gov.cn
印刷：北京利丰雅高长城印刷有限公司
发行：中国林业出版社
电话：（010）8314 3518
版次：2017 年 9 月第 2 版
印次：2017 年 9 月第 1 次
开本：1/16
印张：20
字数：150 千字
定价：280.00 元

酒店是旅人的临时居所。十八世纪以前，应将其统称为“客栈”，是只能提供食宿的供过路人歇脚的地方。然而伴随着经济的发展、物质的丰富，如今人们对酒店的需求已不再只是简单的功能满足了，人们更重视体验，更注重精神上的享受。而近十来年新起的具有独特风格的新型酒店更代表了部分精英旅行者的个人品位和生活水准，甚至已逐渐成为体现城市风貌的时代名片。作为设计师，面对当今越来越高的要求和挑战，要做出好的设计，绝非易事。

何为好的设计？以人为本。似乎这四个字的诠释略为笼统，却是设计的真谛。酒店的以人为本意味着全方位满足客人的各种功能需求和情感体验。设计人性化，功能多元化，透过细节的雕琢提升使用舒适度。同时打造酒店的独特气质，渲染酒店的文化氛围，让客人获得美的观感的同时也感受到一种融入，一种不知不觉中被酒店散发的气场感染并俘虏的深切体验。设计的最高境界就是达成思想与空间的共鸣，从而引发人的思考或是唤起一种情绪。这种情绪可能是祥和的、温暖的、愉悦的、轻松的、甚至充满魅惑的……这取决于设计师的表达。要做到这种空间与情感上的交流互动，需要设计师将自己对世界的认识与感悟凝结在设计中，为酒店注入灵魂，传达给入住的客人，使其获得非凡的居住体验。

当然，这种对酒店文化内核的构塑并不是容易的事。酒店的设计涵盖了建筑规划、园林景观、室内装修、功能布局、流程规划、经营分析等等，设计师的创作需要在各种限制条件中完成，时时面临问题和多方面的挑战，并不能完全实现最初的构想。而成功的设计，必须因地制宜，全局掌控，通过演绎新的创意，打破困局，独树一帜，使结构和文化元素和谐融合。这绝对是一场对创意和智慧的考验，设计师必须在其中找到巧妙的平衡点和切入点，才能使其得以完美的实现。另外，一个优秀的设计师除了经验的厚实累积，更要热爱生活、时时感受、体会生活，吸纳各种文化精髓，密切关注当下前沿的流行元素和趋势，并融合于自己的思想之中，这非常重要。

现在酒店丛书泛滥，像本书这样用心去收录这么多全球具独特个性又精致时尚的酒店实属少见。相信各位读者朋友阅后一定收获良多，并期待日后会涌现越来越多的出色的酒店设计丛书，为社会带来不同以往的文化浸润和独特体验。我们拭目以待。

2012.6.8

目录

设计酒店 DESIGN HOTEL

Contents

设计酒店

DESIGN HOTEL

杭州天伦精品酒店

TEABOUTIQUE HOTEL HANGZHOU

设计 陈奕文

设计单位：杭州天澜设计院
项目地点：杭州曙光路124号
建筑面积：3500 m²
主要材料：混泥土、实木
摄 影 师：陈奕文

杭州天伦精品酒店身处素有天堂之称的历史文化古城杭州，酒店位于黄龙洞景区正对面，身处黄龙商圈核心地段，紧邻黄龙体育中心，美丽的西子湖近在咫尺。

浓郁的茶文化元素和温馨典雅的江南气息淋漓尽致地融入了43间别具匠心的精致客房，每一间客房，为您营造一个自然、放松、恬静的休憩之所。以私密为核心，设有门禁系统，让你感受宁静和回家般温馨。每一间客房都配有西湖双绝：龙井茶和虎跑泉，这是来自杭州的最高礼遇。

天伦精品酒店T SPA推出非常地道的泰式SPA，在一个充满异域风情的环境里接受神态祥和，双手合十向你问候的泰国技师的Massage，尽情享受度假的惬意和新鲜感，让您五觉感受抚慰、身心感受惬意。

步入杭州第一家风格独特的酒廊，一处新颖于原始氛围结合的圣地。“薄荷”融入调酒艺术的本源：品味畅饮鸡尾酒的乐趣。薄荷是一处享受的乐园：伴随着流动的音乐旋律以及深情的节拍，让您可以敞开心扉畅谈的完美音乐花园。户外小憩，薄荷环绕的私语花园是隐藏于闹市之中的宁静。

杭州天伦精品酒店，传统的江南底蕴融合了现代化元素，让“精致江南”和“现代商务”融为一体。

Hangzhou Tea Boutique Hotel is located in Hangzhou which is known as "Heaven on Earth" and famous for its historical and cultural sites. Our hotel is located on 124 Shuguang road, opposite the Yellow dragon cave scenic area, in the center of Huanglong CBD, close to Huanglong stadium, and the West lake.

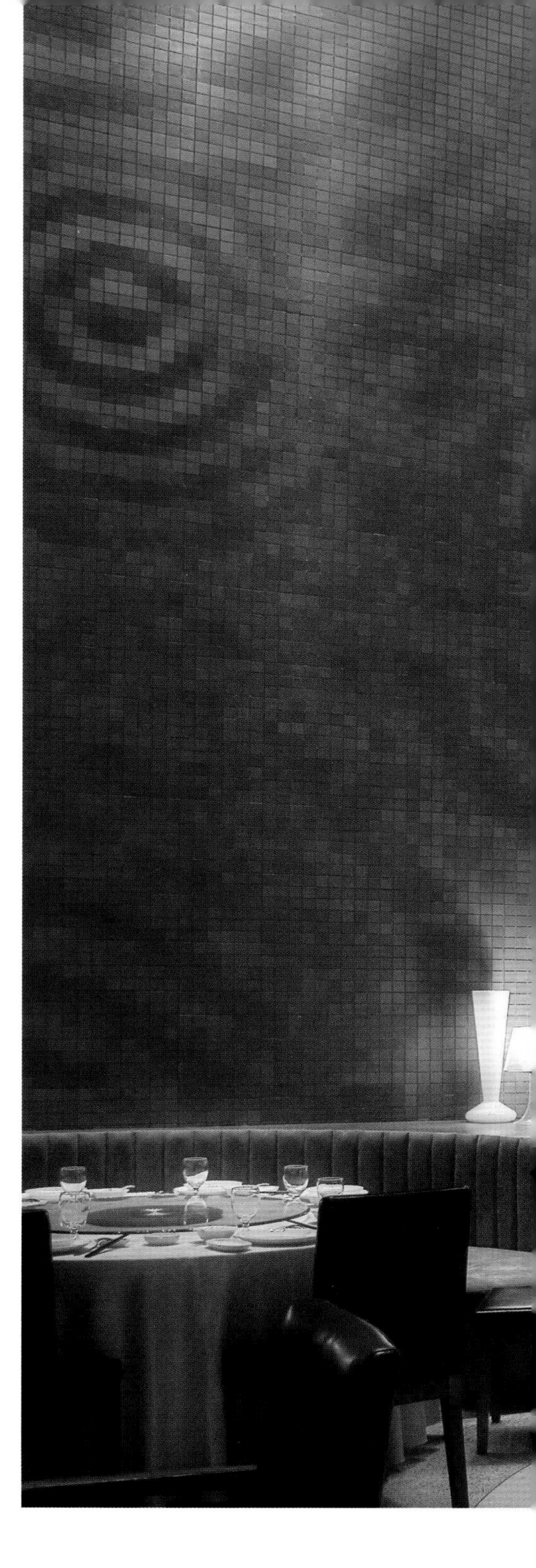

Strong tea culture and elegant chinese traditional flavor is incorporated in the 43 boutique rooms; There are Entrance Guard at the entrance, like your own apartment, customers have to use special card to enter the hall. Tea boutique hotel encourage "low carbon" life style like lower energy consumption, low carbon discharge. This them is not only reflected at the hardware and decoration of the hotel, but also the usage of hotel facilities, Tea is the national drink, and Hangzhou is the tea capital", hangzhou is famous for the West lake and Dragon well tea since ancient times. The Dragon well tea features green color, delicate fragrance, sweet flavor and beautiful appearance and have a long history.

Listening the sounds of nature; Looking the exotic charm; Tasting flavor of flowers; Touching angle's hand; Feeling the heart comforble,Completely relax your body and heart with essential oil in T SPA.

Step in the first lounge of its kind, a sanctuary of unique flavors mixed in original combinations. MINT brings the art of blending into its essential purpose:the pleasure of sipping cocktails. MINT is the residence of enjoyment:a perfect musical garden to share conversations while a flow of lounge melodies let your soul diving into the understanding of deep house music , nujazz and soulful beats. Unique cocktails our lounge's signature.

fashionable and graceful lounge and Thai style SPA make the tradition to meet the modern, blend the "boutique JiangNan with Modern Business Life".

201

Canal House 精品酒店

CANAL HOUSE BOUTIQUE HOTEL

参与设计：Rob Wagemans、
Menno Baas、
Sofie Ruytenberg、
Erik Van Dillen、
Chrissie Phillips
客　　户：Curious Hotels
项目地点：荷兰阿姆斯特丹皇帝运河148号
建筑面积：1225 m²
完工时间：2011
摄 影 师：Jim Ellam、Amy Murrell

Canal House酒店的理念，是建成一个豪华的精品酒店，这里提供了所有五星级可以提供给你的服务，另外还给你带来家庭般的热情和温暖。其宗旨是给宾客提供一个真正的家外之家，在这里，仿佛觉得到了自己的小窝一样。在私人客房旁边，宾客可以根据自己的需要和心情，在会客厅、酒吧、花园，甚至厨房里停留。

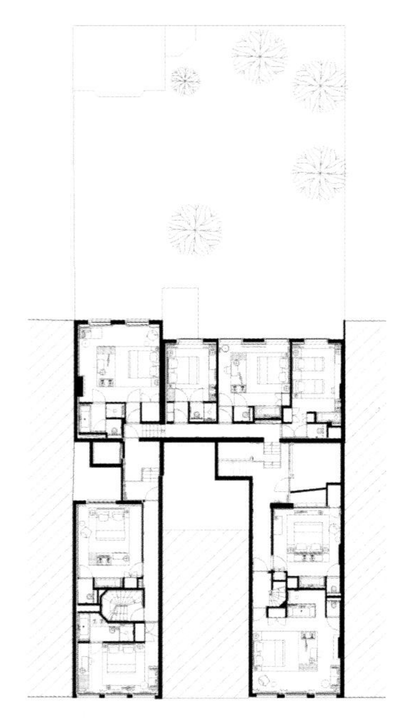

1e verd.

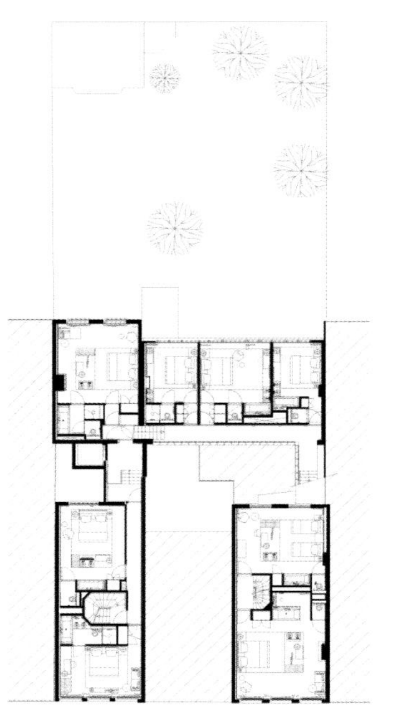

2e verd.

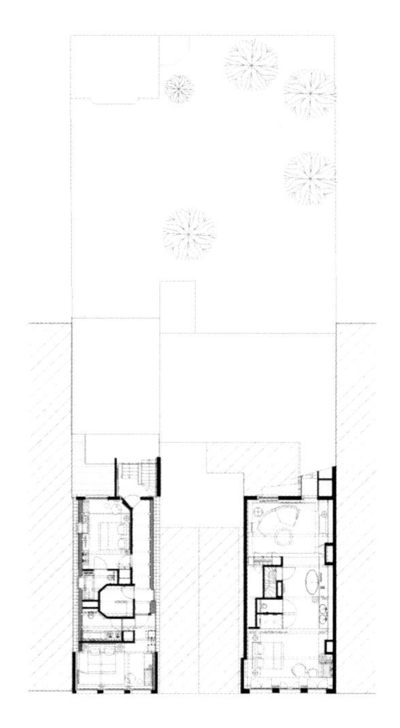

3e verd.

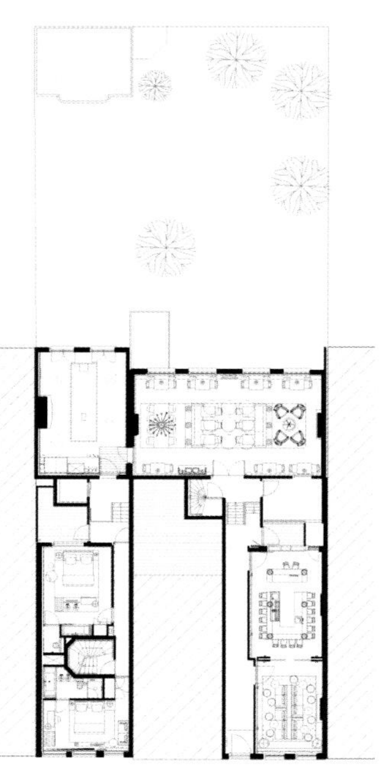

bel étage

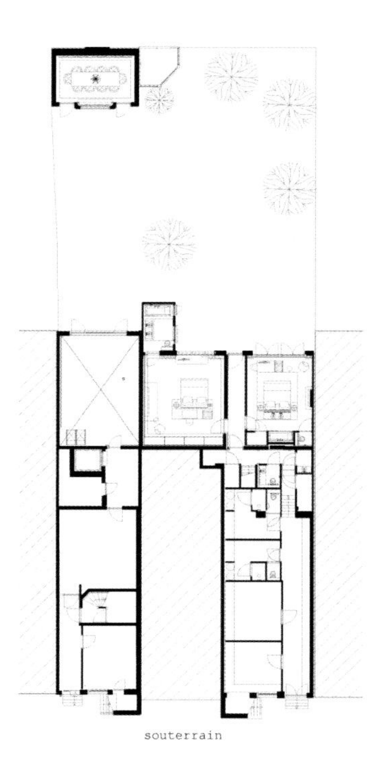

souterrain

Canal House酒店坐落在阿姆斯特丹中心，那里大部分都是怀旧的装饰，地板的饰面和烟囱依旧存在。为了强调这一理念，并与旧Canal House酒店的格调保持一致，所有的装饰品都漆成了黑色。这可以保持它所具有的旧时代的特色，同时又营造了一个当代的环境。

这种抽象的装饰在房间中也是可见的。古老的水罐和碗已经被重新改造。通过放置一个浴缸，浴室和卧室合并到一体，使得卧室显得更宽敞。“卧室”和“浴室”仅由一面镜子分隔开。卫生间位于房间的背面，可以从“浴室”走过去，以保证其隐私。

The concept of the Canal House has been to build a luxury boutique hotel which offers all the service you expect from a five star plus hotel combined with the welcoming and warm feeling you get from a home. The aim was to create a true home away from home, in a house which you could use nearly as if it would be your own. Next to the individual rooms the guests have the opportunity to stay in the living room, the parlor room with a house bar, the garden and even in the kitchen according to their needs and mood.

The Canal House is situated in an old Canal House in the centre of Amsterdam, in which most of the old ornaments, floor finishes and chimneys still exist. In order to underline the concept and therefore to keep the character of an old Canal House in a contemporary way, all ornaments were kept but painted black. This made it possible to remain all the opulence of the old times but nevertheless create a contemporary environment.

This abstraction of the ornament is also visible in the rooms. The water jar and bowl placed of ancient times has been reinvented. Bathroom and bedroom merge in to one united experience by placing the bathtub and the double vanity in the bedroom. 'Bed-zone' and 'bathzone' are only separated by a full height mirror. The toilet cubicle is located in the back-wall of the room and is accessible from the 'bath-zone' to guarantee privacy.

The Club 精品酒店

THE CLUB BOUTIQUE HOTEL

设计 Ministry of Design

参与设计：Colin Seah、
Kevin Leong、
Joyce Van Saane、
Cheryl Lum、
Don Castaneda、
Bryan Law、
Danis Sie、
RobertoRivera、
Lolleth Alejandro
项目地点：新加坡
建筑面积：1950 m^2
摄　　影：CI&A Photography

The Club是设计部最新的高端设计精品酒店，位于保护区客纳街，酒店共设有22间独特的客房，此外还包括一楼的休闲小吃吧、地中海露天餐厅以及屋顶天空酒吧。

在Club中，设计部通过将空间设计、空间附属设计以及招牌设计融为一体，从而编写出一套完整的设计构想。以设计和生活方式为设计目标，The Club集前卫和舒适为一体，它走在全球设计的前列。

设计总监Colin Seah这样说道："我们试图将酒店设计置于新加坡以及客纳街和安详山丰富的殖民历史这两大背景中，我们从这两个方面汲取灵感。"

"首先是新加坡的殖民史，我们已经通过艺术的方式，比如莱佛士爵士的雕像比真人还要大，还有一些主要的家具

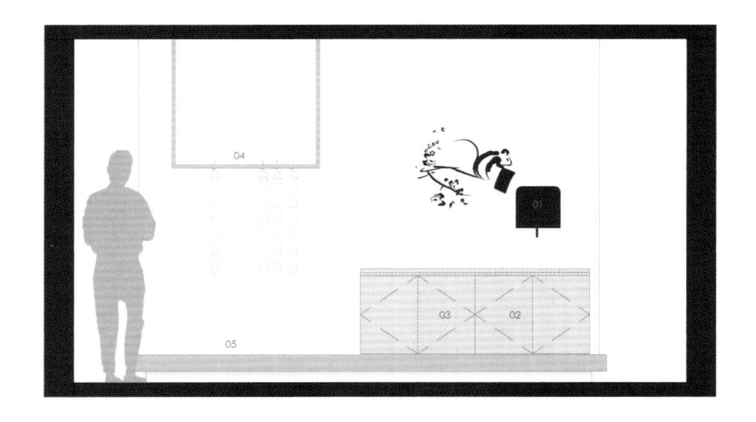

01 WELCOME TRAY 03 SAFE 05 LUGGAGE
02 MINIBAR 04 CLOTHES

THE CLUB TYPICAL MINIBAR COUNTER AND CLOTHES STORAGE

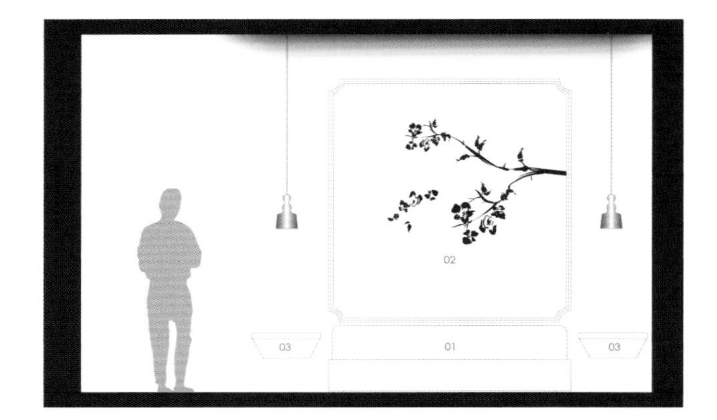

01 BED 02 FEATURE WALL 03 BEDSIDE TABLE

THE CLUB TYPICAL FEATURE BEDHEAD

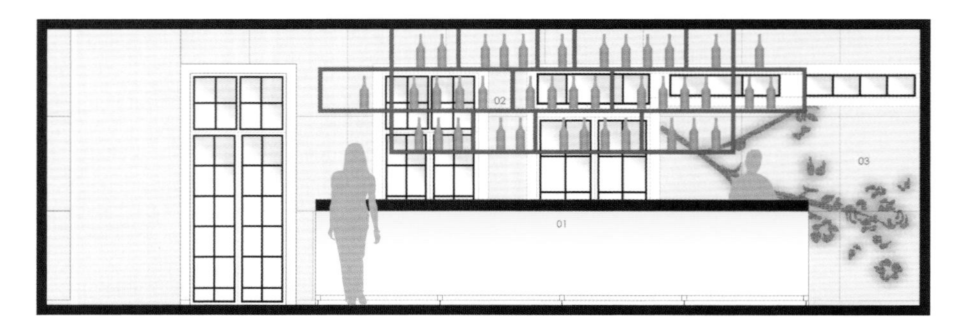

01 BAR COUNTER 02 DISPLAY RACK 03 LED WALL ARTWORK

THE CLUB SKYBAR INTERIOR

01 COUNTER WITH BAR STOOL 02 PLANTER WITH SEATING 03 LOUNGE SEATING

THE CLUB SKYBAR EXTERIOR

和艺术品等，达到了时尚与自然”。

第二个灵感来自该地区的知名度，它是本世纪华人的邮政中心，中国移民在这片土地上辛勤劳作，将一封封渴望的信和一笔笔血汗钱寄回家乡。我们捕捉了历史的遗留记忆和痕迹，在这里，现今的流浪者和过往的流浪者在某一刻实现了交汇。

所有客房都结合了殖民历史中的传统设计元素以及现代井然有序的设计风格——充满了殖民时期特色的美。每个房间中独特的布局以及特殊剪裁的艺术品使22间客房各自呈现出不同的设计风格。设计部设计这些艺术品，且当地著名的艺术家Wynlyn Tan将这些设计融入进该酒店。

酒店在一楼大堂和天空酒吧均设有登记处，从屋顶的天空酒吧鸟瞰，人们可一览客纳街和中心商务区的独特美景。此外，由Jane Yeo 设计的地中海露天餐厅还包括大堂休闲区、小吃吧及两个独立的功能分区。

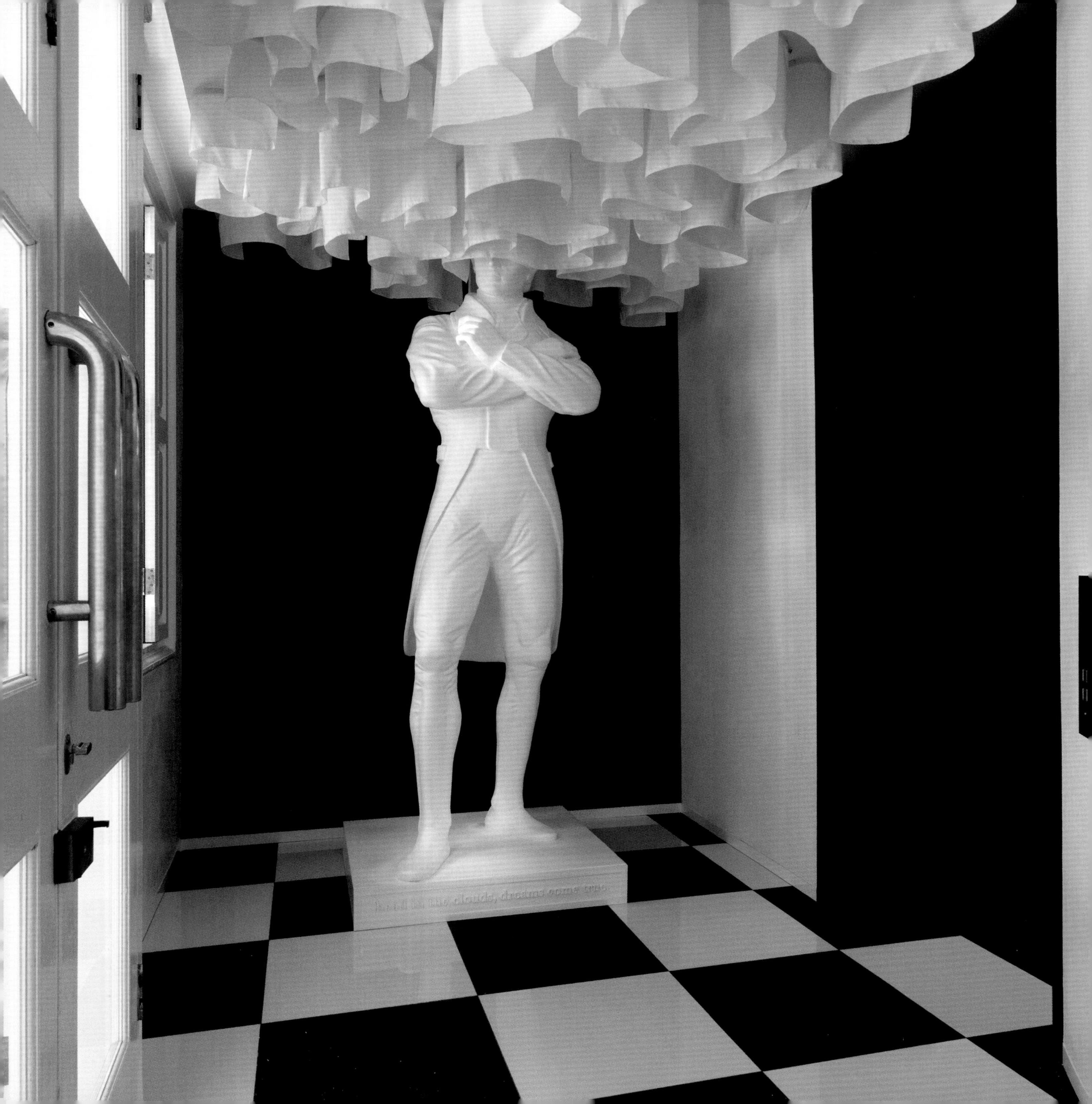

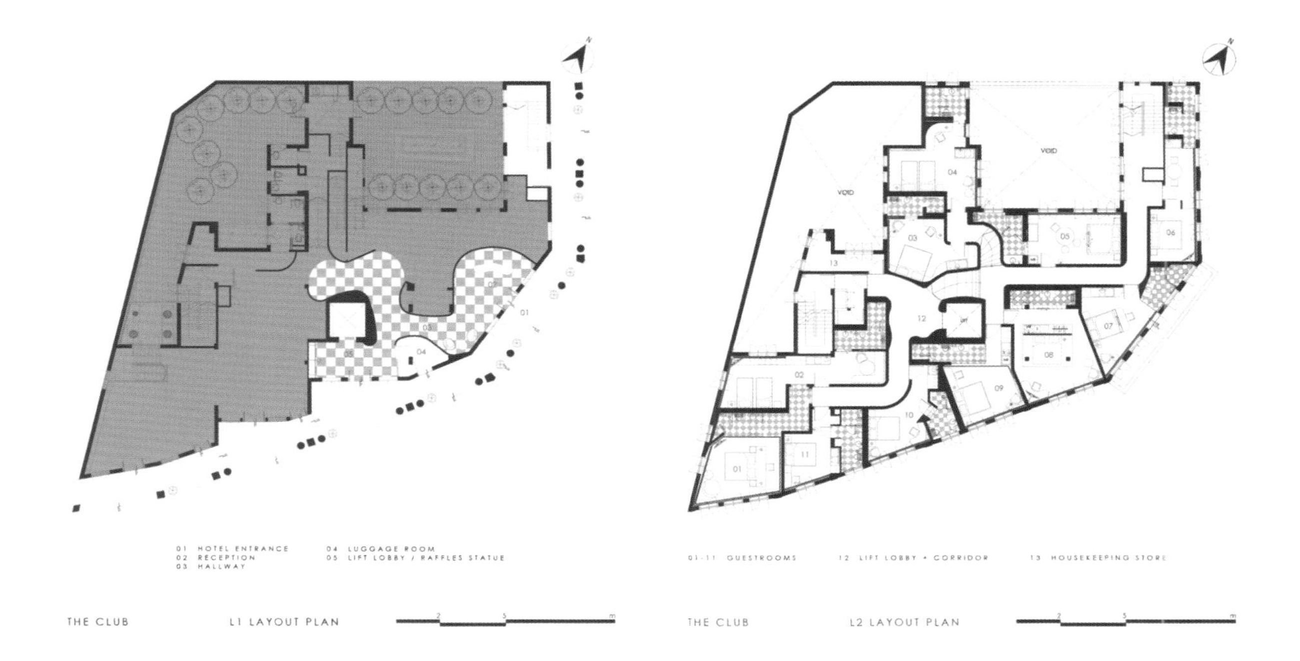

The Club is Ministry of Design's latest high design boutique hotel offering in the uber chic Club Street conservation area with 22 distinctly unique rooms, a rooftop skybar with alfresco deck and a destination F&B venue with a tapas bar on the ground floor.

Conceptualizing The Club's branding, MOD has orchestrated a unified design vision to all related collateral, signage and spatial environments. Targeted at the design and lifestyle savvy global nomad, the Club's blend of sophisticated and comfortable design is at once distinctly local as it is cutting edge global.

Colin Seah, Design Director says, "Searching to ground the hotel in the context of Singapore as well as the historically rich conservation area of Club Street and Ann Siang Hill, we drew its inspiration from 2 sources."

"The first is Singapore's colonial past, which we have made modern tongue-in-cheek references to through art installation like features such as an larger-than-life statue of Raffles with his head in the clouds as well as through some key furniture pieces

and artifacts.

The second inspiration was drawn from the area's popularity as a remittance center for turn of the century Chinese immigrants where hard earned money and wistful letters were sent back to the homeland. We have taken the memories of these exchanges and created features that hint of this legacy in the rooms of The Club, where the modern day nomad and the nomad of yesterday cross paths for a moment."

All rooms combine traditional colonial design inspired elements together with sleek modern detailing, attitude and creature comforts - creating a colonial chic aesthetic. Unique layouts together with tailored artwork in each room make each of the 22 rooms distinct. MOD designed the artwork and famed local artist Wynlyn Tan implemented them in the hotel.

Guests have the option of checking in at the ground level lobby or at the panoramic roof top Sky Bar, overlooking the Club Street conservation area and CBD. F&B areas designed by Jane Yeo include Lobby Lounge, Tapas Bar, and 2 private function rooms.

208
i need

SingTel
OCBC Bank

台北W饭店
TAIPEI W HOTEL

设计　Terry McGinnity

设计单位:G.A. Design International Ltd.
项目地点:台北
摄 影 师:Ralf Tooten

台北W饭店楼高31层，由伦敦著名建筑事务所G.A. Design International Ltd.倾力打造，周身采用玻璃外观，是城内唯一能纵览信义区繁华街道全景的酒店，亦可一睹以逾500 m身量直耸云霄的地标建筑、全球最高楼之一101大厦之芳容。

来到台北W饭店，宾客们定将对这个由坚固不锈钢及镜面设计构成的The Chain(锁链)惊叹不已，它象征了这间酒店对台北的热情与厚爱。此外，一面巨大的绿色墙壁上种满了来自台湾的各种植被，令人耳目一新。步入其中，宾客将开启一段体验台北W饭店独特魅力的全感官之旅。台北W饭店彰显了多种领先的设计布局，将感官享受与空间设计完美结合起来。

互动灯光设施可与走进酒店的宾客互动，随着环境的变化呈现不同的形状、外观及感觉。出自W酒店未来设计师

rAndom Internationa之手的幻光魅影(To Light You Fade)位于台北W饭店的地下层，以一面再生木头墙为背景，宾客可与灯光肆意共舞。该设施配有特制软件，由数百支独特的OLED（有机发光二极管)组成，它们个个出身名门，均来自德国亚琛的全球首条生产线。头顶上方是一个由塑料与竹筒构成的精美雕塑，倒映在镜面之中，施施然由大厅天花板悬垂下来；而“Purple Target”则是一幅镶满优质图钉的帆布，迎接着来到信义中心地带的宾客。安装在入口处及十楼地板上的白色投射灯给人以水雾迷蒙的感觉，走过这里便是独具一格的W Living Room，可以欣赏到WET泳池摩天台尽头的标志性泡沫雕塑。

走进W Living Room，在实木板材堆栈而成的迎宾台前，铺有一幅用红色电子花束点缀的地毯。Living Room摆放着宽敞的圆形睡椅，一扇扇实木百叶窗组成了错落有致的檐篷，若将檐篷折低，便合围而成二楼的会议区。Living Room侧面装有一个11 m高的“W”型屏幕，对面是一个同等高度的壁炉，为您在台北的寒冷夜晚提供一个时尚温馨的消遣去处。宾客还可在由木条拼接的高大茧状座椅营造的私密空间中放松自己，6 m长的大理石吧台将为您随时提供各式果汁及饮料。W Living Room内里有

一整面墙完全由玻璃构成，形成内外通透视野，穿过光滑的玻璃嵌板便可直达WET泳池摩天台。

台北W饭店的特色WOOBAR（与酒店的迎宾前厅并列）的地板以奢华的天然突尼斯石材铺就，取名为“Autumn Brown”（“秋色棕影”），造型独特的圆形白色真皮宽大软凳一字排开，与入口地灯的水滴设计及泳池的水花飞溅遥相呼应。WOOBAR层高10 m，顶棚可随时间进行调整，大厅一端是一座高科技壁炉，另一端则是DJ台，世界顶级DJ将透过操控这些与设计浑然一体的先进音响及灯光系统，给眼光独到的泡吧一族最畅快的享受。WET泳池摩天台清洁明亮，四周环以木栈道，设有开放式壁炉，绿植葱郁，更修建了生态绿墙。泳池一端是一座醒目的金属泡沫雕塑，与银色水珠交相辉映，一群形态各异的陶制蝴蝶振翅欲飞。

10楼的the kitchen table餐厅采用明亮轻快的花园别墅设计，并添加一丝现代气息，内部装修以鲜明的嫩黄色与清爽的白色为主，犹如清晨的阳光洒满房间。the kitchen table在设计之时考虑到了台北的亚热带气候，两面墙壁均安装了超高滑动门，直通毗邻的WET泳池摩天台露天就餐区。奇幻的雕塑顶棚就像一个藤蔓盘绕的篱架，装饰简约的

混凝土柜台又仿佛一张张公园长凳。互动的开放式厨房中央是一个2.7 m长定制Moletini炉灶。the kitchen table的其它墙面均以书架覆盖，上面摆放着装满阳光的罐子——每一个罐子都是一支太阳能LED灯泡，白天可搬到WET泳池摩天台上充电，晚上再运回餐厅在架子上发光。

位于台北W饭店的最高层31层的紫艳中餐厅，提供新广东料理，亦是W的全球首间中式餐厅。紫艳中餐厅的落地大窗将信义区的繁华街景一览无遗，夜色中的霓虹灯流一直蔓延至台北周遭的远处山脉。紫艳中餐厅展示取材自中国传统烹饪工具的原创艺术品，创意大胆前卫。有一个大型墙面装置是由数百个金属汤勺组成，排成立体圆圈；另一个引人瞩目的装置则是由中式瓷制汤匙组成，造型错综复杂。数千个形状各异、大小不一的饼干模具以独特造型组成两个并列装置，而餐厅侧面则是一组由创意十足的中式厨具组成的墙面雕塑。

紫艳中餐厅墙面以深紫色亮漆与光亮的望加西黑檀结合，隐匿在夜色之中，烘托出无限美景。厨房中央是一个迭层厨房，外镶黑色大理石，每当夜幕降临，烹饪区便忙碌起来。座位区摆放宽敞，排列为云层状，周围环绕由粗绳织就的高达5 m的屏风。匠心独运的屏风，恰似中国的鸟笼，与之遥相呼应的是不锈钢餐台上高挂的丝线灯笼。酒吧偏居于31层的一隅，由玻璃幕墙围绕而成，堪称台北最气派的场所之一。酒吧外墙由一面面多棱镜组成，在双层高的紫水晶屋中闪闪发光，给食客带来新鲜的用餐体验。一套专为台北W饭店打造的"飞碟灯"，布满酒吧的屋顶，透过温馨的灯光，宏伟的台北101映入眼帘。

台北W饭店拥有405间客房和套房，使其跻身台北规模最大的奢华饭店之列，每一处私人空间都可欣赏无以伦比的都市美景。暖色调石料、抛光实木、绿意盎然的植物雕花地毯与中式灯笼罩子中充满现代感的精致灯光形成了令人沉醉的奇妙对比。所有客房均设有室内阳台／休闲区，宾客可以在台北这座国际大都会中享受一份属于自己的宁静，350纱支密度亚麻的W特色睡床将为您带来一夜酣眠，优雅现代的白色书桌搭配有符合人体工学的皮椅。浴室配有超大海岛风情浴缸，与以地铁为灵感的红色或黄绿色壁面砖及木质隔断形成对比，让您放松身心。考虑到国际飞行旅客的需求，每间客房均配备了高科技设施，包括：高速有线及无线互联网接入；42英寸平板液晶电视；Bose声道系统；iPod充电座；提供语音邮件服务的IP电话以及为客房增添更多奇妙色彩的W特色十二生肖。不少客房进行了台北式的改造，木墙上悬挂着题为"你在哪里"("Where Are You")的解构地图，窗外一边是阳明山的自然美景，一边是台北的万家灯火。解构地图正是电力十足的台北的缩影，给宾客带来一丝逃脱喧闹的幽静之感。

Taipei W hotel is 31 floors high, which is designed by famous architecture institute of G.A. Design International Ltd. in London. Its entire body is made from glass, which is the only hotel that can enjoy the whole prosperous landscape of Xinyi District. It also allows you to view the 500 meters high landmark, the face of 101 Building which is one of the highest building in the world.

Entering Taipei W hotel, visitors will surely surprise for the Chain consisting of stainless steel and mirror design. It symbolizes the passion and love for Taipei of this hotel. Furthermore, there are "planting" various of Taiwan plants on a huge green wall, which is much fresh to people. Enter it, visitor will have a great senses experience of Taipei W hotel's special charming.

The interactive light device coupled with the walking people, shows different type, appearance and feeling with the changing environment. To Light You Fade created by rAndom Internationa of W hotel is located at the underground. Coupled with a recycled wooden background, visitors can dance with light easily. This device is equipped with special soft ware. It is consisted with hundreds of unique OLED (organic light emitting diode). They are all well-produced from the world's first production line of Aachen, Germany. On the head, there is an exquisite sculpture of plastic and bamboo. Reflecting on the mirror, they look like hanging from the ceiling. Furthermore, Purple Targe is a canvas studded with high-quality pushipin to meet the guests of Xinyi Cen-

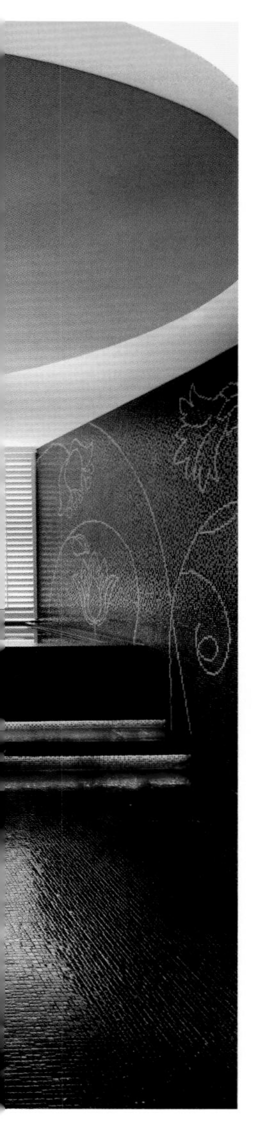

ter. The white spotlight installed at the entrance and the 10th floor give us a dreamy sense. Passing through it, there is a unique and special W Living Room, where you can enjoy the WET swimming pool platform' s bubble sculpture.

Entering W Living Room, you will see the wooden reception table and the carpet paved with red electronic bouquet pattern. There is placed a huge round couch, wooden shutters in the Living Room. They are placed in orderly. If put the canopies lower, then they will form a conference area at the second floor. At the side part of the Living Room, there is installed a W type screen with 11 meters high. Facing it was a fireplace with same height, which will provide you a fashion and warm place in cold winter. Guests also can relax themselves in the huge cocoon-like seat made of wooden strips. You can enjoy the fruit juices and drinks at the marble bar with 6 meters high the time you like. There is one side of wall in W Living Room made of glass completely, forming a transparent vision. Passing

through the glass panels, you will come to WET swimming pool directly.

Taipei W hotel's unique WOOBAR (in the same line with the reception table of hotel) is paved with the luxury natural Tunisi stone, named as "Autumn Brown". The special round white leather soft chair are placed in order, which is echoing with the drop design at the entrance and the spraying water in the swimming pool. WOOBAR is 10 meters high, which can adjust the roof by time. There is a high-tech fireplace at end of the hall. At the other end, there is a DJ table. The world-class DJ will be played here. Coupled with the advanced stereo and the light, you will have a great vision experience. WET swimming pool is clear and bright, surrounding with wooden road. There is set an open fireplace. Seeing the flourish trees, it is like entering a ecology park. At one end of the swimming pool, there is an shining golden bubble sculpture, echoing with the silver water drip, looking like a group of butterflies.

The kitchen table at tenth floor is adopted bright and lovely garden villa design. Added with some modern art sense, the internal decoration is adopted the bright yellow and white as the theme, which seems like the morning sunshine. When designing the kitchen table, considering the subtropical climate of Taipei, the two walls are all installed with ultra-high sliding doors, straight to the WET swimming pool platform for food. The magical sculpture roof is like trellis vines. The simply concrete counter is like benched in park. At the interactive open kitchen central is a 2.7 meters long Moletini stove. The other walls of the kitchen table are covered by bookshelves. There are placed many sunshine jars—each jar is a solar LED light bulbs. During the daytime, they can be moved to the WET swimming platform for charging and then take back at night.

The Purple Chinese Restaurant located at 31the top floor of this hotel provides Guangdong cuisine, which is also the first Chinese restaurant of W. The large windows of Purple Chinese Restaurant have absorbed the bustling street of Xinyi District completely. The neon lights have spread to the distant mountains around Taipei. The exhibition materials of Purple Chinese Restaurant come from Chinese traditional cooking tools, with boldness creation. There is a large wall made of metal spoon and arranged in three-dimensional circle. The other eye-catching device is made of Chinese porcelain spoon. Thousands of various shapes and sizes biscuit molds are made into two parallel devices in special shape. The other side of the restaurant is a group of creative Chinese kitchen wall sculpture.

The wall of Purple Chinese Restaurant is made of purple lacquer and bright Garcia ebony, hidden ii the night, seeming much beautiful. The central of kitchen is a laminated kitchen, which is inlaid black marble outside. When night falls, the cooking area will be busy. The seats are placed in a spacious way. They are surrounded by five meters screen made of rope. This excellent screen likes a birdcage in China. Echoing with it is a silk lantern hanging above stainless steel dining table. The bar is located at a corner of the 31th floor, surrounding with glass walls, which can be called as the most impressive place. The outside wall of bar is made of many prism mirrors, shinning among the double height amethyst house. It will bring a fresh dining experience to visitors. A set "UFO" lights made for Taipei W hotel covers the whole roof of the bar, passing through the warm light, the grand Taipei 101 will come to your eyes.

The Taipei W hotel possesses 405 rooms and suits, making it be listed in the largest luxury restaurant in Taipei. Each private space allows you to enjoy the beauty of this urban. The warm color stone, polished wooden, green plant sculpture and the Chinese lantern cover are all full of modern art sense, letting people immerse into such nice atmosphere. All the rooms are equipped with balcony/leisure area. The guests can enjoy a piece of quiet of themselves here. 350 pieces of thread-count linen W beds will bring you a refreshing night. The elegant modern white desk is coupled with manmade leather chairs. The bathroom is equipped with a large island-style bathtub. They have formed a contrast with the subway-inspired red or yellow-green wall as well as wood partition, allowing you have a fully relaxation. Considering of the needs of international flight passengers, each room is equipped with high-tech facilities, including: high-speed wired and wireless internet access, 42-inch flat screen LCD TV, Bose-channel system, iPod cradle, the IP phone with voice mail service as well as the wonderful funny Zodiac. Many rooms have been transformed in Taipei style. There is a deconstructed map named "Where Are You" hanging on the wall. Seeing through the window, there is a beautiful Yangming Mountain and houses of families. This deconstructed map is just the microcosm of Taipei, bringing a sense of peace to visitors.

安缦法云
A MAN FAYUN

设计　Jaya Ibrahim

设计单位：Jaya & Associates
项目地点：杭州
建筑面积：140000 m^2
主要材料：石材、砖、土木

安缦法云位于西湖西侧的山谷之间，距杭州市中心20分钟车程。沿路两旁竹林密布、草木青翠，经过植物园和西湖内部的支流溪涧，便来到天竺寺和天竺古村落。安缦法云即坐落于天竺古村另一侧。这里共有47处居所，始建于唐朝，曾为附近茶园村民居住。垣墙周庭，充满自然之趣，宛如传统中国村落的缩影。

安缦法云的接待总台就掩映在绿荫翠竹之中，由此沿一条幽径即可通往度假酒店的主干道——法云径。法云径连接所有客房（庭院住宅）和酒店设施。这里的住宅可追溯至百年以前，如今以传统工艺修缮一新，砖墙瓦顶，辅以土木结构，屋内走道和地板均为石材铺置。

法云径总长600 m，通往酒店餐厅、村庄食坊、茶室、精品店和水疗中心。酒店东侧有一条小溪由南而北缓缓流过，它曾经是古村落日常生活的聚集地，村民们在茶园辛勤劳作了一天，日近傍晚便汇集于此，临溪沐浴，闲聊畅谈。

Located in the valley on the west side of the West Lake, Amanfayun Hotel is twenty-minute drive from the downtown of Hangzhou City. Along the either side of the drive there are dense bamboos, green grass and trees; Tian-zhu Temple and the Relic of Tian-zhu Village are beyond the Botanical Garden and tributaries that flow into the West lake. Amanfayun is on the other side of the Relic of Tian-zhu Village. There are totally 47 houses here and their construction were started in Tang Dynasty and they used to be the homes of the villagers planting tea nearby. The walls, the houses and the courts are full of natural taste, as if it was a miniature of a traditional Chinese village.

The reception desk of Amanfayun is in a patch of green bamboos, where a lonely path may lead you to the major route of the resort, Fayun Route, which connects all the guestrooms (houses in the court) and the facilities of the hotel. The houses here may trace their history back to one hun-

dred years ago and they have been renovated with traditional technology, such as bricked walls, tiled roof, structures of earth and wood, all the floors inside houses paved with flat blocks of stone. The total length of Fayun Route is 600 meters, leading to the restaurant, eating house in the village, tea house, boutique and water therapist’ s. On the east side of the hotel there is a stream flowing slowly from the south to the north, which used to be the meeting place of the old village in daily life. After hard work in the tea garden, the villager gathered here at dusk, washing themselves in the stream while talking and laughing.

嘉兴月河客栈

JIAXING YUHE INN

设计　陈向京　曾莹

设计单位：广州集美组室内设计工程有限公司
参与设计：皇甫丹琳、肖正恒
项目地点：嘉兴市
建筑面积：23000 m^2
主要材料：金丝柚、黑胡桃木、伯利黄、山西黑
摄 影 师：罗文翰

三面临水，八面来风。嘉兴作为江南文化的发源地，自古为富庶繁华之地。早在六千余年以前，先民便在此繁衍生息，孕育出璀璨的马家浜文化。遂历经“滨海泽国”、“嘉禾飘香”、“繁庶市镇”、“文风鼎盛”等重要历史时期。“滨海泽国”——水乡泽国，河网密布，是嘉兴自然环境的最贴切写照。“嘉禾飘香”——沃土嘉禾，为漕粮供应之地，稻作文化兴盛。“繁庶市镇”——明清时期，纺织业发达，市镇星罗棋布，实为丝绸之府。“文风鼎盛”——文人雅仕问鼎一时，风流人物传盛一世。经时代传承，这些历史时期的民风意象部分已升华为当地的民俗精粹，演绎成一抹别样的江南风情，成为嘉兴最吸引人之处。

在月河客栈的设计中，我们通过对当地建构手法的研究与继承，传统装饰纹样的抽象简化以及本土化材料的解析运用，有机地将禾香、庶市、文风、泽国，这些极具民风意向的主题融入室内空间，揉古释今，化凡为雅，营造出极具江南水乡风情并富有现代气息的主题体验空间。

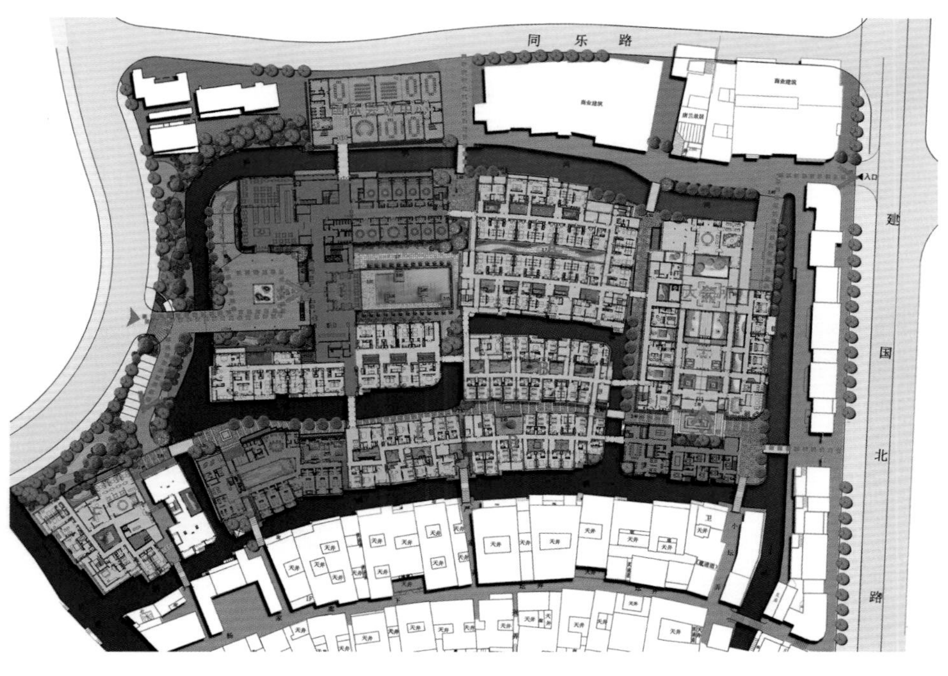

Surrounding by waters in three directiong, as the motherland of Jiangnan culture, Jiaxing is a flourishing city since ancient times. Six thousand years ago, people began live here and generated the famous Majiabang Culture. Then it has experienced "Coastal kindom", "Jiahe fragrance", "Flourishing town", "Great Culture" and other important historical period. "Coastal kindom"--waterside place, with internet rivers, is the most closly description of Jiaxing. "Jiahe fragrance"--fortile land of Jiahe, is a grain place, particularly of rice. "Flourishing town"--in Ming and Qing Dynasty, the spinning industry is flurish, fulfilled the town, which can be called as Home of Silk. "Great Culture"--here was full of talents and literati. After heritage from generation to generation, these customs and culture of history have been improved to the essences, performing a unique Jiangnan scenery and becoming the most attractive place of Jiaxing.

In the design of Yuhe Inn, through the study and inheritance of the local construction, simplification of the traditional decorative patterns as well as the utilization of the local materials, we have merged plant seeds, ancient market, cultural, water into this space. Combined ancient and modern together, converted general to elegance, we have created a most Jiangnan mood scenery with modern atmosphere.

27

玖

锦宅

JING'S RESIDENCE

设计 Antonio Ochoa Piccardo

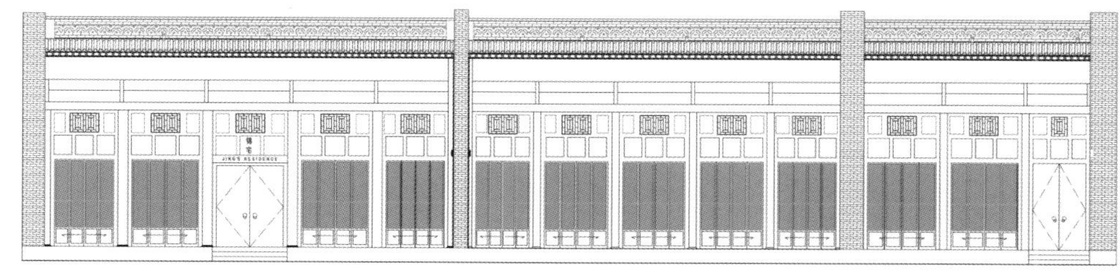

side street elevation

设计单位：安东红坊建筑设计咨询（北京）有限公司
项目地点：山西平遥
建筑面积：1600 m²

位于清代建筑内的特色酒店，采用极少的设计手法，以期强调古建筑的原有美感。

锦宅特色酒店位于中国最古老的三座城市之一的平遥古城，那里至今仍维系着古香古色的建筑群落，锦宅特色酒店主要由两个老院落组成，两个院子在建筑外观上都延续了传统特色，而在室内设计中融入了新的设计元素，将中国传统文化与现代生活做巧妙的诠释。锦宅的设计思想是：旧建筑的室内外设计是在展现历史风貌和体现现代生活方式中寻求设计契合点，在古今结合中寻求出设计的契机，设计师用极少的设计手法，以强调古建筑的原有美感。建筑伴随着时间的流逝而愈有质感，安静祥和；院中的竹子却是新生命的开始，跃跃欲试蓬勃不息。

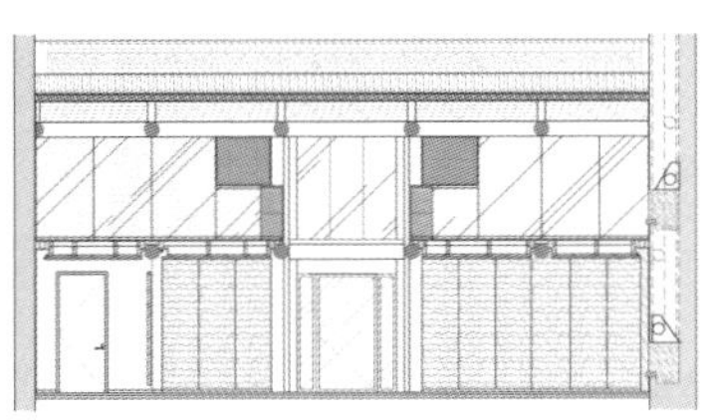

Antehall&Bar elevation 1:50

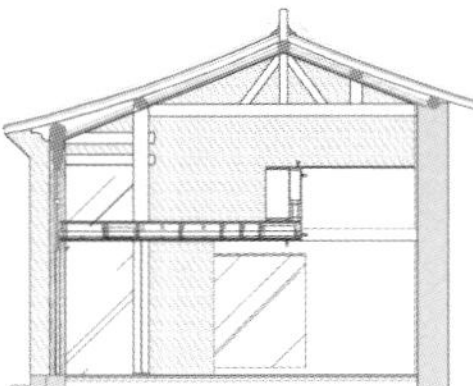

Antehall&Bar elevation 1:50

Antehall&Bar elevation 1:50

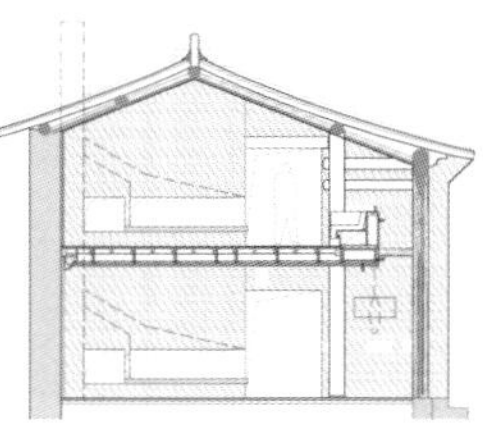

Antehall&Bar elevation 1:50

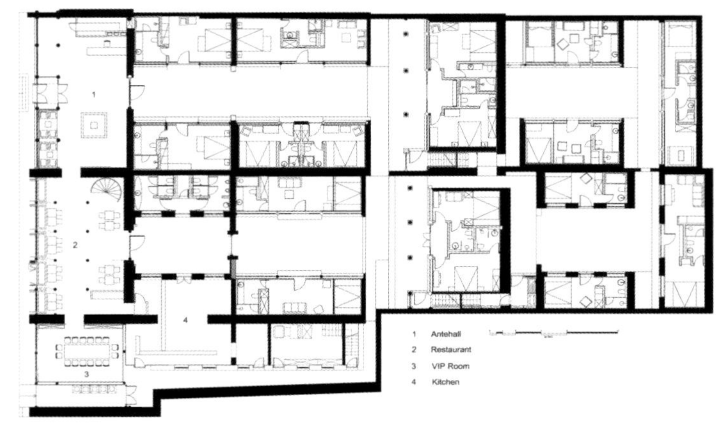

1F plan & landscape 1:200

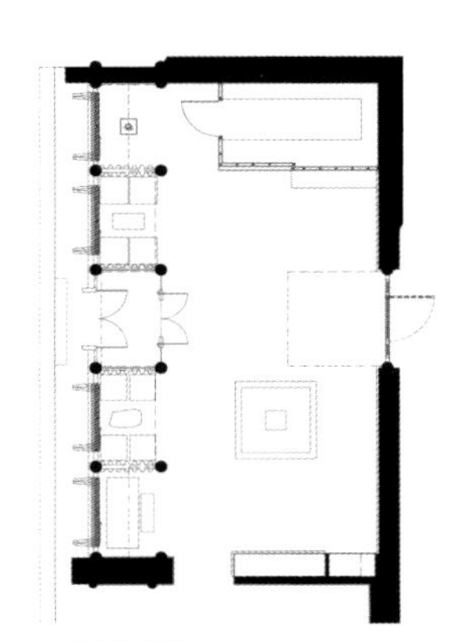

Antehall Plan 1:100

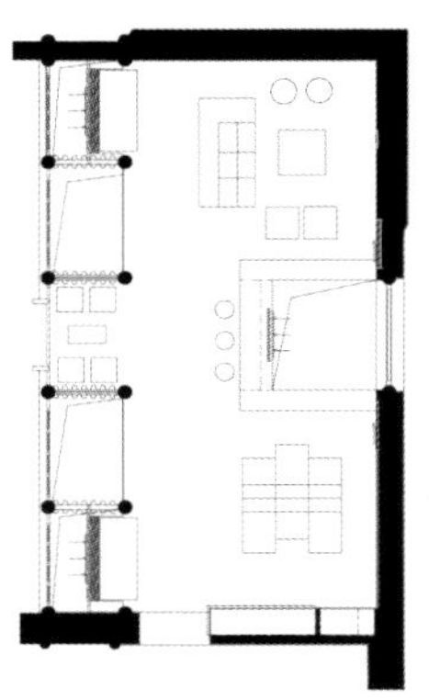

Bar Plan 1:100

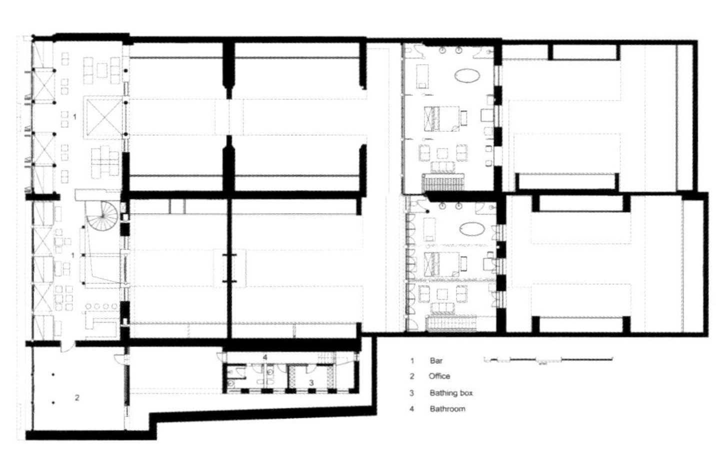

2F plan 1:200

This hotel, locating at architecture of Qing Dynasty, is adopted a few design ways. It is mainly to emphasize the original beauty of the ancient buildings. Jinzhai Unique Hotel is located at ancient city of Pingyao, which is one of the three oldest cities. There are many ancient buildings even today. The Jinzhai Unique Hotel is consist with two old courtyards. These two courtyards are all keeping the traditional characteristics. However, in the design of interior, there are many new design elements, which has demonstrated the Chinese traditional culture and modern life fully. Design concept: the design on the old buildings of Jinzhai is showing the history and modern life style, which is seeking the common features by combining the ancient and modern style. Adopting a few design manners, the designers is mainly to stress the original beauty of the ancient building itself. Buildings are more charming with the time passing, quietly but peaceful. The bamboo in the courtyard is indeed the beginning of life, flourish and lively.

裸心谷
PURE HEART VALLEY

设计　Delphine Yip

首席主规划师及建筑师：benwood Studio Shanghai
夯土小屋建筑师：A00 Architecture
园境建筑师：Design Land Collaborative (DLC)
环保单位：Environmental Resources Management (ERM)
可持续性：安生态有限公司 (BEE)
LEED：上海太平洋能源中心 (SPEC)
室内设计单位：AIM
室内空间规划：benwood Studio Shanghai

裸心谷位于浙江省风景秀丽的田园胜地莫干山，这个强调可持续理念的豪华养生中心兼牧马自然保护区占地67英亩(约27万 m^2)，坐落于一个私人山谷之中，四周环绕着大型水库、翠竹、茶林以及一些小村庄。度假村内拥有121间客房，分布于独栋树顶别墅以及夯土小屋之中——所有这些建筑全部采用业内领先的可持续材料建造。度假村内的用餐选择包括拥有80个座位，位于水库旁边的中非合璧餐厅Kikaboni(在非洲语中意为“有机”)、泳池酒吧兼西餐厅Clubhouse Cafe和24小时的客房送餐服务。度假村内拥有占地750 m^2的养生水疗中心naked Leaf，中心的15间理疗室设于竹林内的高脚屋之中，在竹影婆娑下更显宁静安谧。裸心谷更设有一个1000 m^2的会议中心Indaba(在非洲语中意为“首脑集会”)，拥有8间多功能会议厅和两个分别俯瞰着竹林和水库的大型平台。此外还设有一间会所和一间名为Little Shoots的儿童游乐兼托儿中心。

naked Galleries是针对追求文化内涵的休闲或商务旅客而设，位于会所往

上的4栋风格独特的建筑环绕着一个小湖，每栋建筑都拥有特色文化主题：茶亭、竹艺馆、项目馆及陶艺室。宾客可以尝试制作富有地方特色的手工艺品、品尝裸心谷出产的“白茶”、在陶艺转盘上捏制一个茶壶、亲手采摘新鲜的茶叶，体验当地的主要生活方式、竹林种植，以及了解裸心谷与当地农民合作的众多环境保护项目。坐落于中央的圆形剧场将会举办音乐会及文化表演等节目。

树顶别墅和融合亚非风格的室内设计均出自benwood Studio Shanghai的知名建筑师Delphine Yip-Horsfield之手。benwood Studio Shanghai负责的知名大型项目包括集餐饮和娱乐于一体的上海新天地等。夯土小屋则是由上海本土环保建筑设计师A00 Architects操刀。匠心独运的室内外设计，令秀丽的山野景观更显迷人，带来返璞归真的自然体验，同时最小化对环境带来的影响。

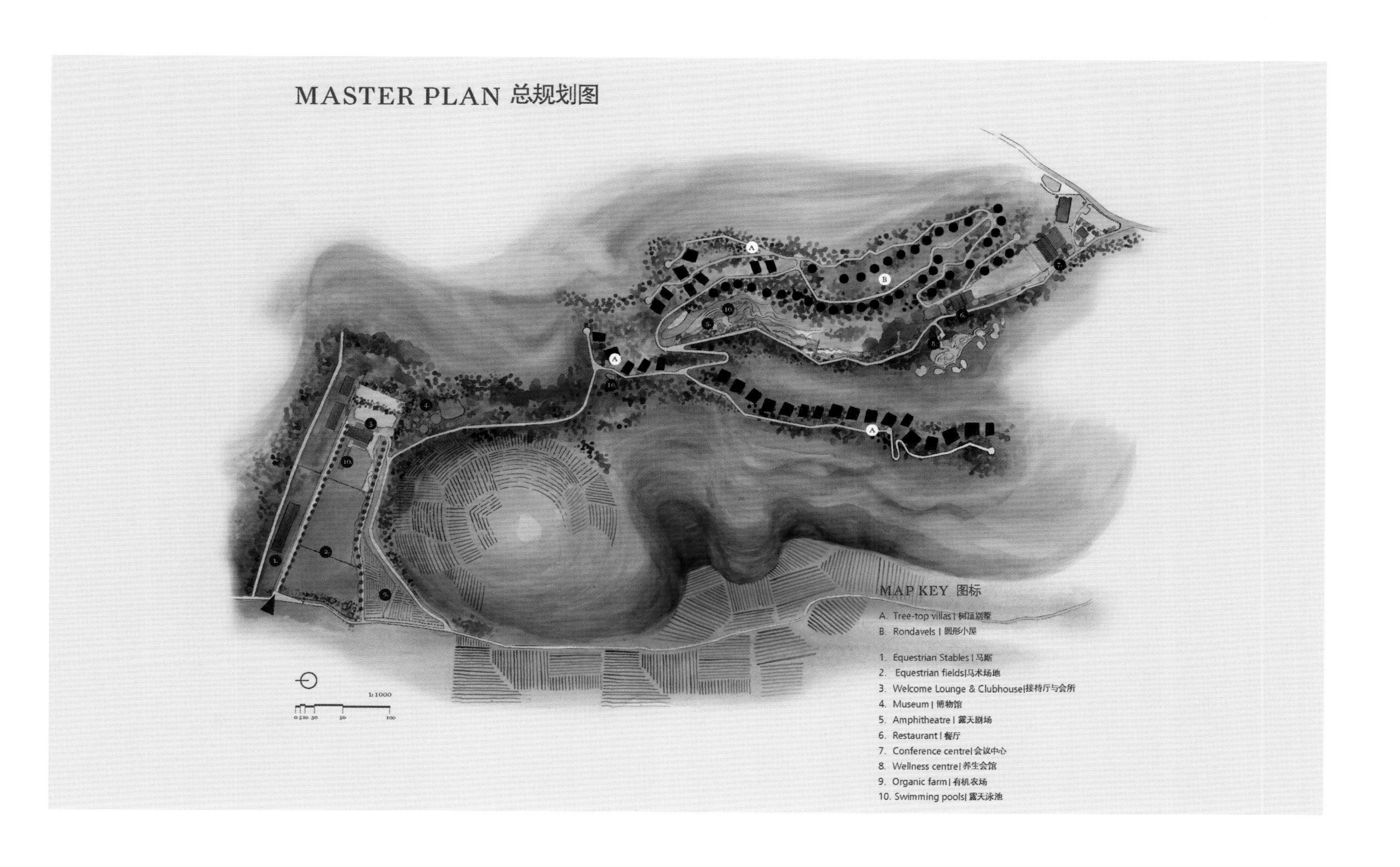
MASTER PLAN 总规划图
1:1000
0 5 10 20 50 100
MAP KEY 图标
A. Tree-top villas丨树顶别墅
B. Rondavels丨圆形小屋
1. Equestrian Stables丨马厩
2. Equestrian fields丨马术场地
3. Welcome Lounge & Clubhouse丨接待厅与会所
4. Museum丨博物馆
5. Amphitheatre丨露天剧场
6. Restaurant丨餐厅
7. Conference centre丨会议中心
8. Wellness centre丨养生会馆
9. Organic farm丨有机农场
10. Swimming pools丨露天泳池

30栋树顶别墅均采用两层设计，提供豪华的双卧室、三卧室或四卧室套房选择。高挑的楼层和从地面到天花板的玻璃幕墙带来无障碍的绝佳视野。别墅的第二层高于树顶，室内为设备齐全的厨房和客厅，而开阔的室外露台设有浴缸、烧烤台和用餐区。所有房间和别墅均配有高速无线上网、36寸卫星电视、CD机、DVD机、iPod基座、高级音响和中央空调系统。水疗式淋浴、纯天然卫浴用品、独家床品和随传随到的管家服务，为您送上极致奢华的享受。7栋特别设计的贵宾树顶别墅拥有独一无二的迷人景观，提供更周到的服务和配套设施。

Pure Heart Valley is located at beautiful Mogan Mountain, Zhejiang Province. This luxury health center is pursuing of sustainable development concept, covering 67 hectares. It is also a Wrangler Nature Reserve locating at this private valley, surrounded with huge water pool, bamboos, tea plants and some villages. This hotel possesses 121 rooms, scattering in different independent villas and small ancient houses—all these architecture are all adopted the advanced sustainable materials. The restaurant in this resort possesses Kikaboni (means "organic" in Africa language) with 80 seats that locating beside water pool, Clubhouse Café that has swimming pool, bar and western food, as well as the 24-hour room service, there is a naked Leaf water spa covering 750 square meters. These 15 spa rooms are placed in the bamboos, which seem quieter and peace under the shadow of these bamboos. Pure Heart Valley even has a meeting center Indaba (means "President Conference" in Africa language) covering 1000 square meters. It has 8 functional meeting rooms and two huge platform to enjoy the bamboos and water pools separately. In addition, there is a club and a room named Little Shoots children playing center.

Naked Galleries is set for guests that pursuit of cultural leisure and business function, locating at four unique architectures surrounding a small lake ahead of the club. Each building has its own unique culture theme, Tea House, Bamboo Art Pavilion, Project Room and Porcelain Room. People can try to make the most location handmade arts, tasting the "white tea" generated by Pure Heart Valley, making a tea pot on the porcelain turntable, pick up the fresh tea leave by yourself, experiencing the local life

style, bamboo plant as well as the various environment friendly projects cooperated with the local famers. The round theater locating at the center will hold concert and cultural performance.

The villa located higher than trees and the Sino-Africa indoor design both come from the famous architecture Delphine Yip-Horsfield of benwood Studio Shanghai. The famous huge project that charged by benwood Studio Shanghai includes Xintiandi Shanghai that merges restaurant and entertainment into a single whole. The ancient small house is made by the local architecture A00 Architects. The special indoor and outdoor design making the beautiful mountain more charming, bringing us a natural and true experience. They are all environment friendly.

These 30 villas located higher than trees adopt two floors design, providing luxury two-bedroom, three-bedroom or four-bedroom suits for selection. The high building and the wall glass window has brought a perfect vision. The second floor of the villas is higher than the trees top. They are all well-equipped with kitchen and halls, bathtub, barbecue, and dining area at the outdoor terrace. All rooms and villas are all equipped with wireless internet, 36 inches satellite TV, CD, DVD, ipod Base, high-class stereo and central air conditioning system. Spa shower, pure natural washing products, exclusive bedding materials and the on-time service will provide you a luxury experience. Seven special villas possess unique and charming landscape with most perfect services and equipment.

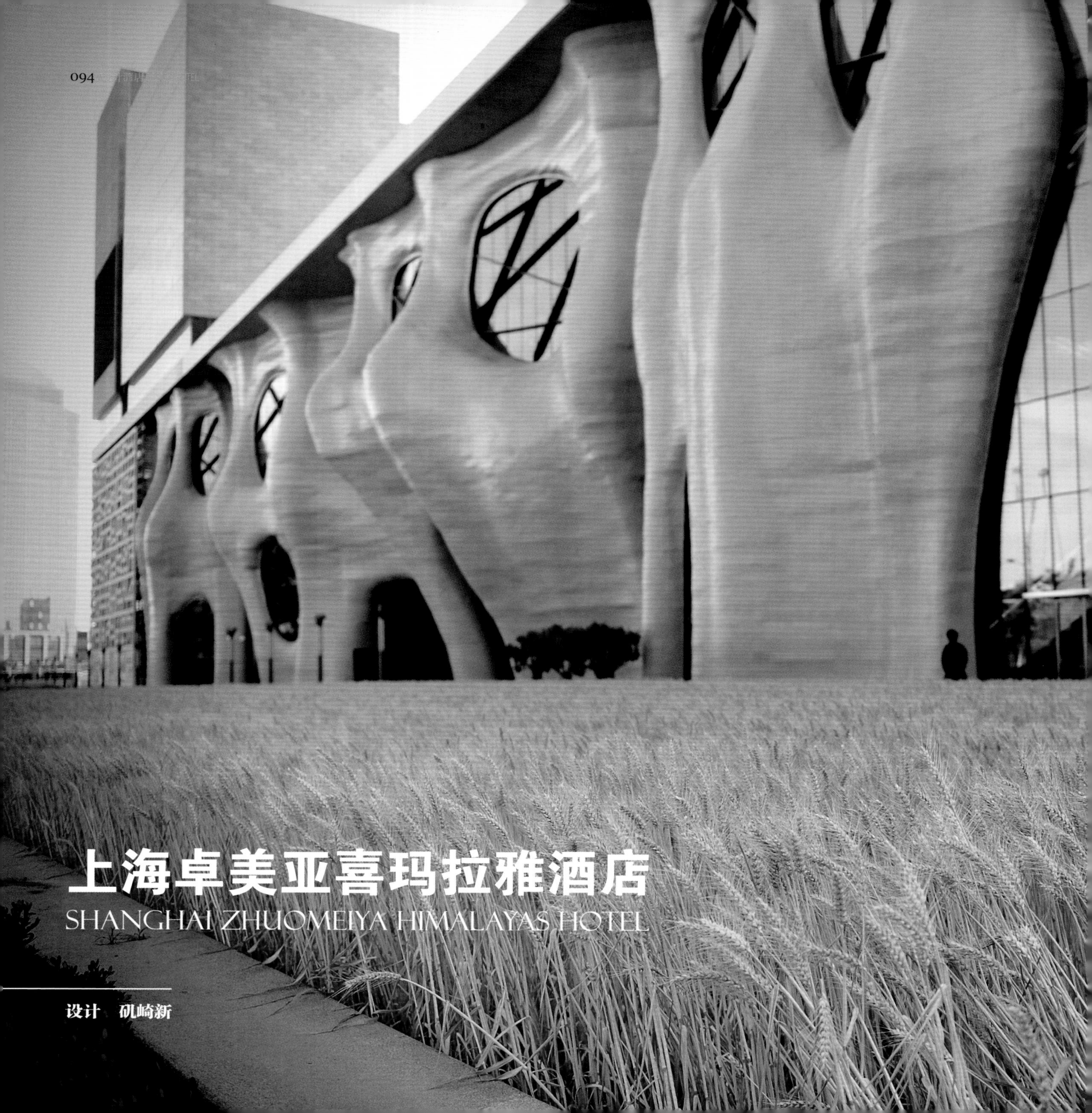

上海卓美亚喜玛拉雅酒店

SHANGHAI ZHUOMEIYA HIMALAYAS HOTEL

设计　矶崎新

设计单位：矶崎新工作室
项目地点：上海浦东新区
梅花路1108号
建筑面积：164549 m²
主要材料：混凝土、挤塑聚苯板、
铝合金玻璃
摄 影 师：朱少慈

上海卓美亚喜玛拉雅酒店隶属的喜玛拉雅中心是上海一座新兴的艺术文化中心，囊括了1100座的大观舞台、喜玛拉雅美术馆、大型购物商场以及5000 m²的屋顶花园。该中心由曾主持设计西班牙巴塞罗那体育馆和洛杉矶现代美术馆的国际著名建筑大师矶崎新先生设计，酒店绝美的内部设计则由曾成功设计卓美亚集团另一地标建筑迪拜帆船酒店的KCA国际倾力打造。其创新的设计理念及非凡的建筑风格灵感来源于中国文化理念和风水原理。

三维立体的异形林犹如参天大树破土而出，诠释了喜玛拉雅中心所要传递的精髓，顶部5000 m²的空中花园连接两座大楼。灵感源于古代树林的根部，融入了风水理念，象征大自然天与地紧密结合。异形林不只作为造型更具备承载空间的结构功能，卓美亚喜玛拉雅酒店的另一酒店出入口也位处异型林之中。

异形林的建筑设计在中国建筑史上没有任何先例可以借鉴。利用真实树干为模本，以内铸钢丝网外附木模的形式，每隔30 cm用电脑进行力学测算，与自然界树木的生长方式一致，并且保证了工程结构的安全。这项工程没有任何标准进行验收，为此，上海市建委单独设立了验收标准，填补了中国异型施工验收标准的空白。

喜玛拉雅中心的两幢大楼由7层楼高的文字字符包围，由黄帝时期的"造字圣人"仓颉创造。它是对古老的文字和中国文化历史的抽象演绎，也是对当代科技的礼赞。每片文字通过特别定制，并且选能工巧匠细心拼贴，矗立出一道充满想象的"文字墙"。底层的橱窗，巧妙的安排间隔，日光由字符间隔内撒进酒店内部。

卓美亚喜玛拉雅酒店楼体结构源自中国古代崇天敬地的礼器"玉琮"形制，内圆外方象征天与地，柱体中贯穿孔高14

个楼层，传达天与地之间的沟通，蕴含中国“天圆地方”的宇宙观，量体安排虚实相应，大度恢宏。柱体底部幽静私密的花园是修身养息的绝佳选择。

酒店大堂内围镶嵌由唐代书法大师怀素书写的千字文。《千字文》最早应梁武帝要求创作，由250个四言短句组成，千字长诗，首尾连贯，音韵谐美，内容有条不紊地介绍了天文、自然、修身养性、人伦道德、地理、历史、农耕、祭祀、园艺、饮食起居等各个方面。

Shanghai Zhuomeiya Himalayas Hotel affiliated to Himalayas Center is a new art culture center of Shanghai, including 1100 seats stage, Himalayas Art Pavillion, large shopping center as well as 5000 square meters roof garden. This center is designed by famous architecture designer Mr. Ji Qixin who has ever designed Span Barcelona Stadium and Los Angeles Modern Art Pavillion. The excellent interior design of this hotel is designed by KCA International which has designed another landmark Dupai Sailing Hotel of Zhuomeiya Goup. Its innovation design concept as well as the excellent architecture style comes from Chinese culture concept and Fengshui theory.

The three-dimension style seems like a huge tree coming out from the earth, demonstrating the soul of Himalayas Center. The 5000 square meters roof garden has connected two buildings. This inspiration comes from the root of ancient forest, merging with Fengshui theory, which is a sign of combining the god and earth. This unique forest is not only used to as shape and also bear the spacial structure function. The other entrance of Zhuomeiya Himalayas Hotel is also set in this unique forest.

The architecture design of this unique forest has no example to copy in Chinese architecture history. Taking teal tree trunk as the mode, adopting the interior steel net as the form, they are calculated by computer every 30

cm. It has the same grow methods with the nature trees and guarantee the safe of the construction. This construction has no acceptance standards. Therefore, Shanghai Municipal Construction Committee established a special acceptance standards for it, which has made up the blank of China's strange building acceptance standards.

The two buildings of Himalayas Center are surrounded by 7-floor high characters, which are created by Cangjie, "Character Maker" in Premier Huangdi. It is also a kind of praise for modern science and technology. Each piece of character are made by special. After jointing them together carefully, they have formed a "Character Wall" which is full of imagination. The windows of the first floor allows sunshine coming into the hotel through these character gaps after carefully design.

Zhuomeiya Himalayas Hotel building structure comes from the shape of "Jade Yuzong" which is a piece of

ware for worshiping god in ancient China. Round inner and square outer shape symbols of the sky and earth. The hole that passes through the column is 14 floors high, conveying the communication between the god and earth. It contains the universe value of Chinese "Round Sky and Square Earth". They are set in reasonable way, seemingly magnificent. The quiet and peach park at the bottom is a best choice for rest.

The inner room wall of the hall is embedded with Tang Dynasty master calligrapher Huai Su's "Thousand-character Article". "Thousand-character Article" was created in the request of Emperor Liang. It consisted with 250 four-sentence poem and thousands of characters long poems. They were coherent and orderly with harmonic rhythm, which had introduced the astronomy, natural, self-cultivation, human relations, ethics, geography, history, farming, worship, gardening, daily life and so on.

水舍

THE WATERHOUSE AT SOUTH BUND BOUTIQUE HOTEL

设计　郭锡恩　胡如珊

设计单位：如恩设计研究院
参与设计：Debby Haepers,
Cai Chun yan,
Markus Stoecklein,
Jane Wang
项目地点：上海
建筑面积：2800 m^2
摄 影 师：Derryck Menere

坐落在上海南外滩老码头新规划区内的水舍，是一座仅有19个客房的四层精品酒店。建筑本身由建造于20世纪30年代三层楼高的日本武装总部改建而成。酒店濒临黄浦江，与闪烁着璀璨灯光的浦东天际线隔江相对。如恩设计研究室（Neri & Hu）对这一建筑的改造设计理念基于"新"与"旧"的融合，原有的混凝土结构被保留还原，大量新加入的耐候钢，仿佛在叙述着这座位于黄浦江边的运输码头的工业背景。而对原建筑进行的第四层加建，不仅与来往于黄浦江上的船舶产生了工业本质上的共鸣，更为此建筑赋予了历史和本土文化的背景。

如恩设计研究室(Neri & Hu)同时还负责该酒店的室内设计。设计师对室内和室外空间，公共与私密空间运用了模糊倒置手法，制造了一种空间迷失感，让那些厌倦了普通五星级酒店、渴望拥有与众不同体验的客人们耳目一新。通过隐避的圆孔和玻璃隔断，公共空间内的客人可以窥视到私密空间（当客房主人敞开私密空间时），同时私密空间客人的视线也被引向公共区域，比如酒店前台上方巨大的垂直窗户以及俯瞰着用餐区的走廊窗户。意想不到的视觉组合在带来惊喜元素的同时，使酒店客人置于一种上海本土的城市状态，一种由视觉走廊和相邻弄堂群所带来的关于上海的独特空间风味。

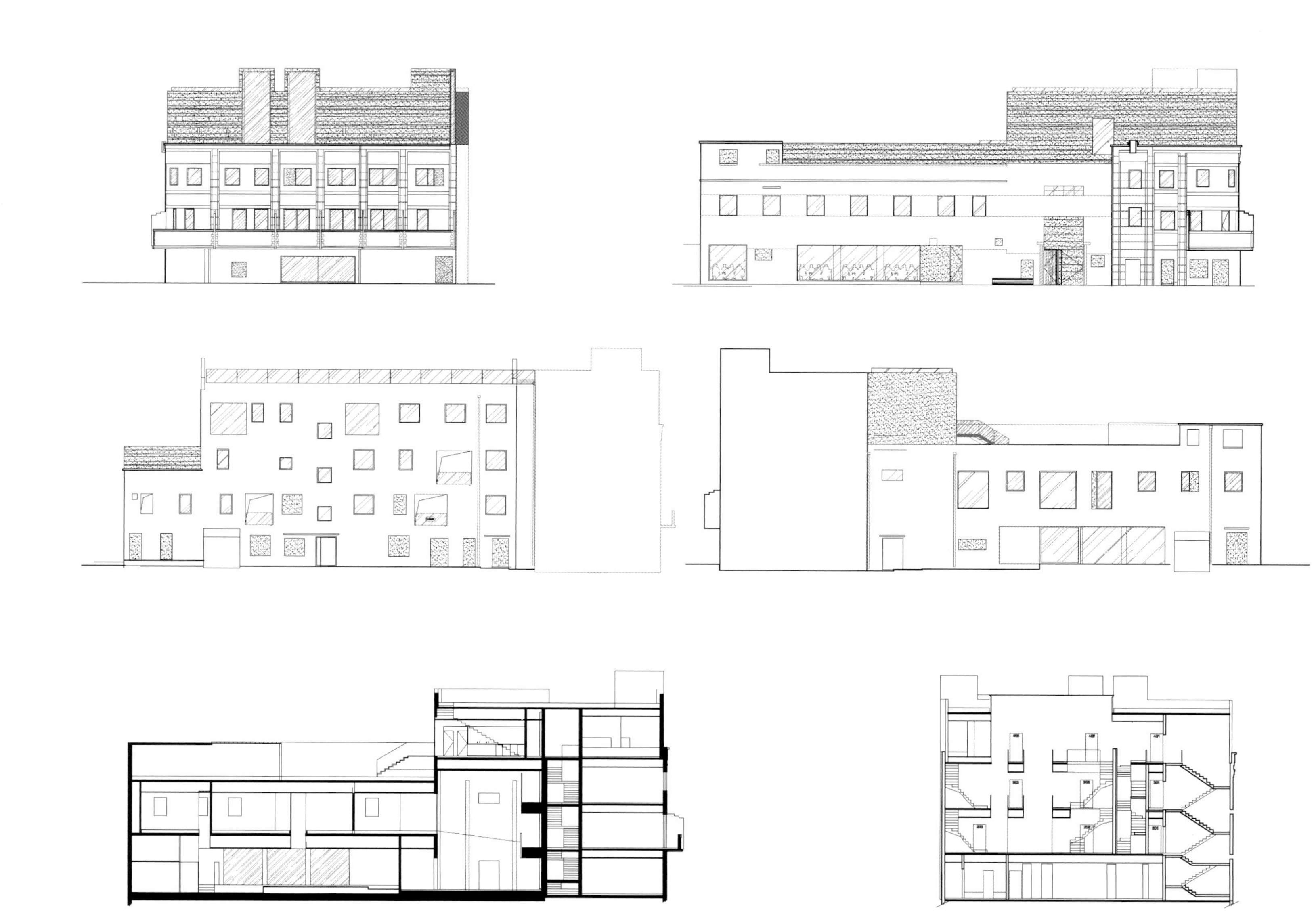

Located by the new Cool Docks development on the South Bund District of Shanghai, the Waterhouse is a four-story, 19-room boutique hotel built into an existing three-story Japanese Army headquarters building from the 1930's. The boutique hotel fronts the Huangpu River and looks across at the gleaming Pudong skyline. The architectural concept behind NHDRO's renovation rests on a clear contrast of what is old and new. The original concrete building has been restored while new additions built over the existing structure were made using Cor-Ten steel, reflecting the industrial past of this working dock by the Huangpu River. Neri & Hu's structural addition, on the fourth floor, resonates with the industrial nature of the ships which pass through the river, providing an analogous contextual link to both history and local culture.

Neri & Hu was also responsible for the design of the hotel's interior, which is expressed through both a blurring and inversion of the interior and exterior, as well as between the public and private realms, creating a disorienting yet refreshing spatial experience for the hotel guest who longs for an unique five-star hospitality experience. The public spaces allow one to peek into private rooms while the private spaces invite one to look out at the public arenas, such as the large vertical room window above the reception desk and the corridor windows overlooking the dining room. These visual connections of unexpected spaces not only bring an element of surprise, but also force the hotel guests to confront the local Shanghai urban condition where visual corridors and adjacencies in tight nong-tang's define the unique spatial flavor of the city.

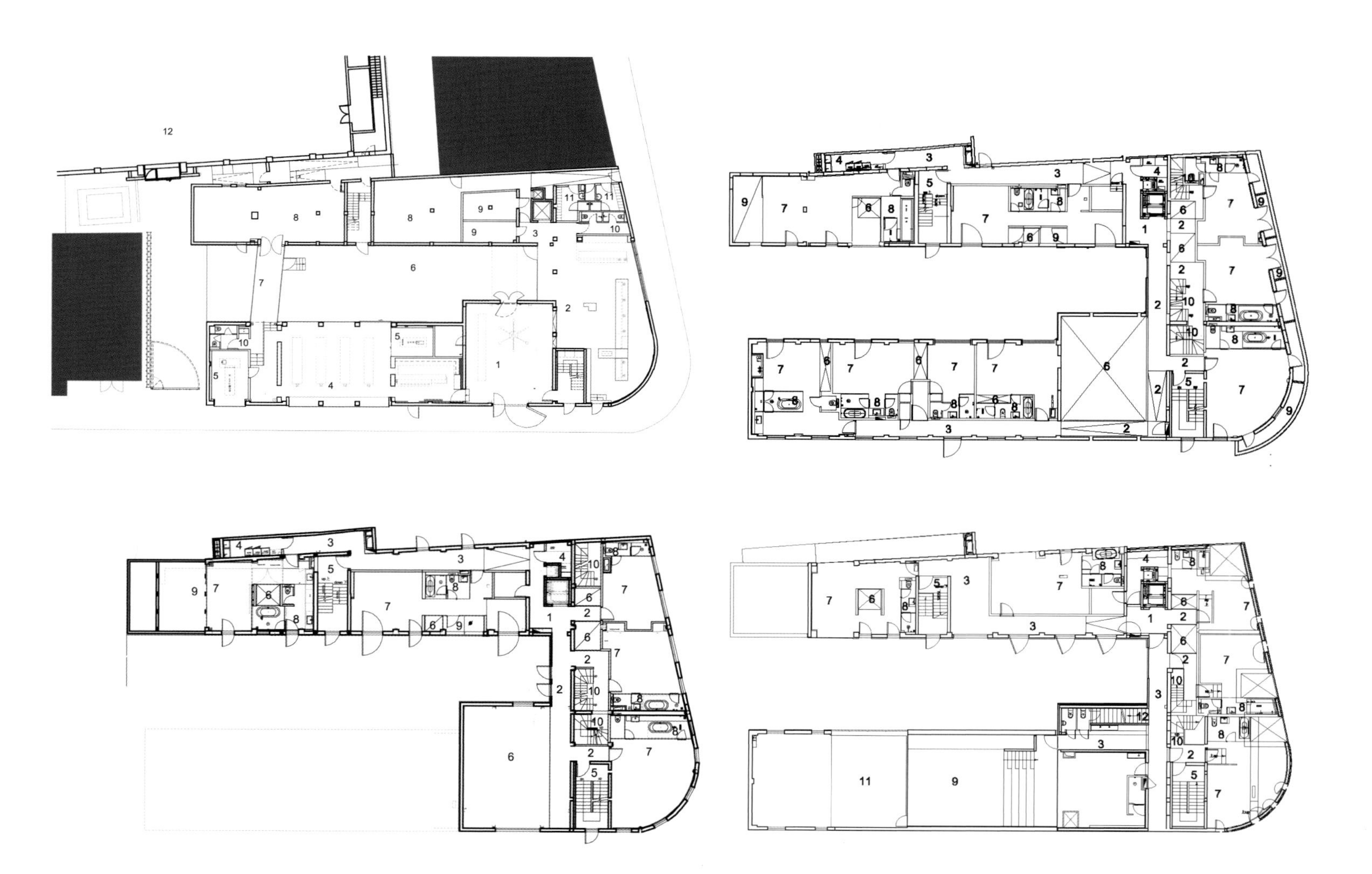

无锡灵山精舍

WUXI LINGSHAN JINGSHE

设计　陆嵘

设计单位：HKGGROUP
参与设计：慎曦、李婷、田珺
项目地点：无锡滨湖区
建筑面积：9800 m²
主要材料：竹、实木、青砖、竹编藤编、墙纸、青灰色凿毛石材、木地板、鹅卵石、图案地毯、透光膜等
摄 影 师：刘其华

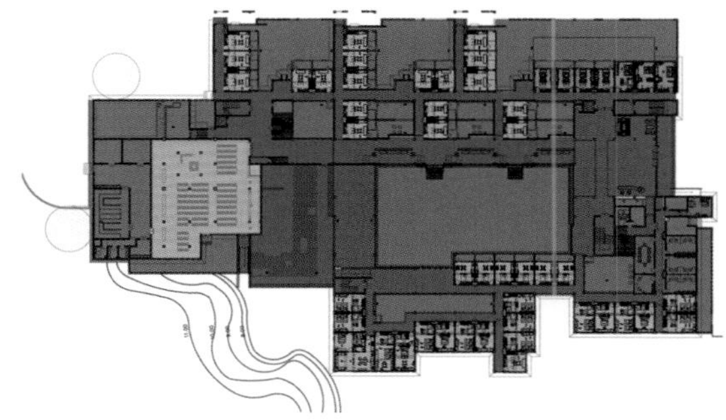

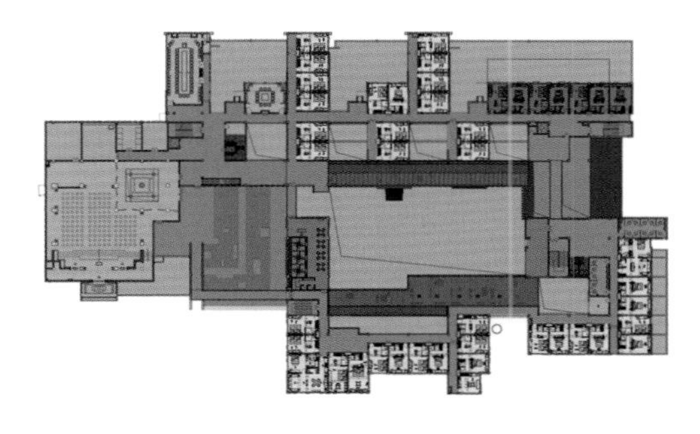

无锡灵山精舍坐落于无锡灵山胜境内，毗邻灵山大佛，总体建筑规模9800 m²左右，拥有90间左右客房，掩映在一片安静的竹林之中。来这里的客人们可以安下心来修身养性、体悟禅境并参加精舍里提供的各项与参禅相关的活动。在精舍的室内设计中，设计师以竹为母题，传承佛陀千年前在印度竹林精舍时的意境。一入大堂，木质的带有风化感的条形格栅天顶、旧铜打造的前台，还有竹子做的几盏大吊灯让人的心一下子就沉静下来。朴素的客房，简单但很精巧，透过细密的竹帘，目光可以穿越到窗后禅意的小院子里。禅堂素雅，竹子的顶棚，竹子的天灯都照应着向禅的心灵回归自然无我。茶室里的家具简约但也渗透出禅境，让客人更好地在参茶的过程中调节心境。这所精致Resort正以“禅”为主题，提供客人“禅”的教诲，“禅”的感悟，“禅”的意和境。

Wuxi Lingshan Jingshe is located at Lingshan Mountain, Wuxi City, closing to Lingshan Buddha. Its overall construction area is about ten thousand square meters, bodering 90 rooms and hiding in this quiet bamboo trees. The guests here can calm down themselves, enjoy the Buddha spirit and participate various Buddha activities provided by Wuxi Lingshan Jingshe. For the internal design of this Jingshe, the design takes bamboo as the main tone, inheriting the Indian bamboo pagodas happened thousand years ago. The time you entering this hall, the wooden with weathered sense zenith, the old copper reception table, and several chandelier made from bamboo will calm your heart and soul suddenly.

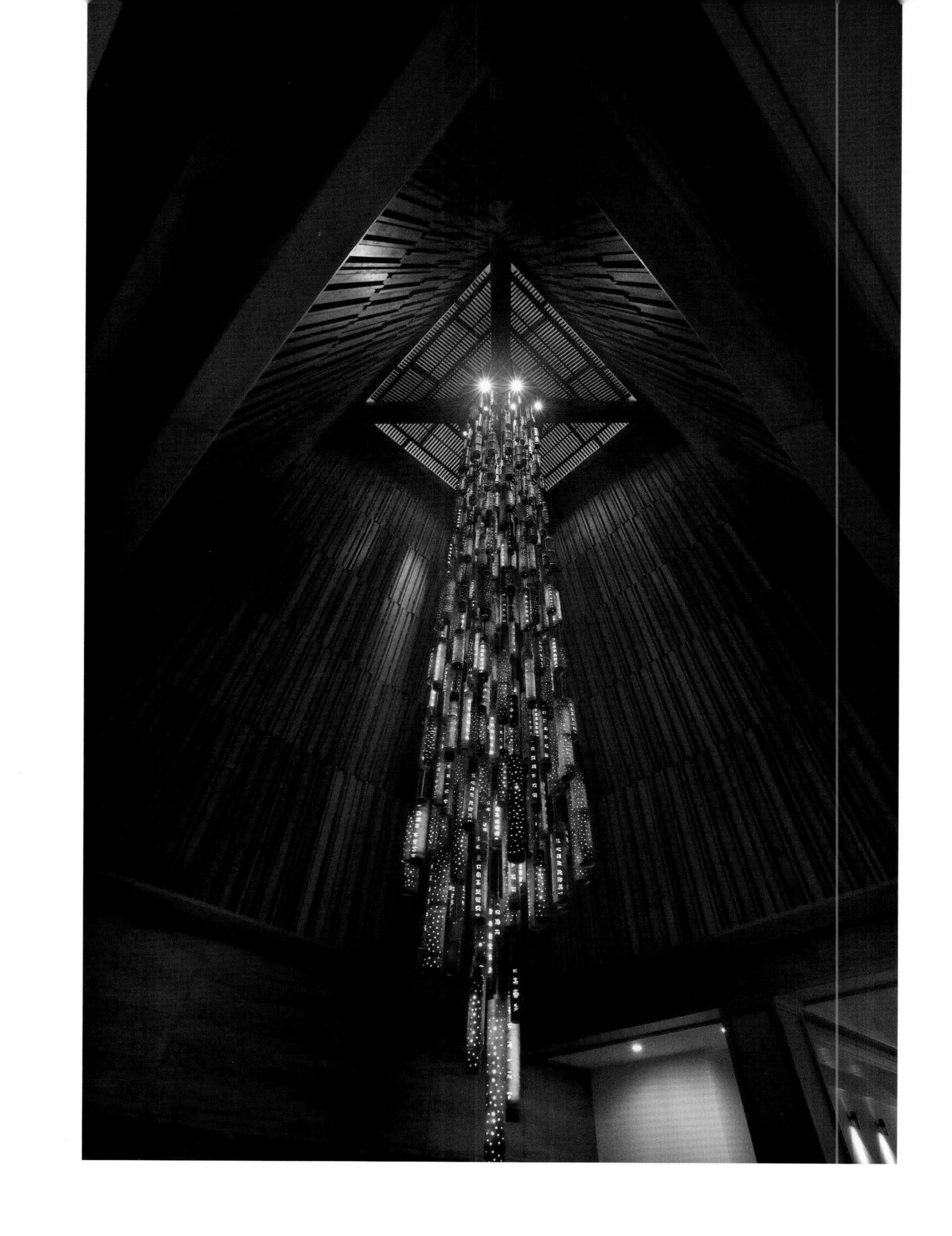

拉萨瑞吉度假酒店
ST. REGIS LHASA RESORT

设计　Jean-Michel Gathy

设计单位：Denniston International Architects & Planners Ltd.
设 计 师：Jean-Michel Gathy
项目地点：拉萨

被古代藏民奉为“诸神之地”的拉萨是西藏的精神与政治中心。坐落于古老的八廓街的拉萨瑞吉度假酒店位置便利，周围尽是各类文化地标，而宾客们则能够在这些充满传奇色彩的建筑物中探索拉萨的终极魅力。这座令人惊叹的度假酒店四周环绕着幽美的自然环境与拉萨那些震撼人心的山脉美景，酒店致力于为西藏带来更高品质的服务水准与奢华享受。富丽堂皇的度假酒店为宾客提供特色瑞吉管家服务、三家融汇各国美食的世界级餐厅、一间著名的品茗轩、一个酒吧以及面积达到 1087 m²尺的 Iridium Spa 水疗中心。这里的泳池装饰华丽，散发着金色的光芒。另外，宾客还能够在沉思园中锻炼瑜伽、进行普拉提训练或者静思冥想。

拉萨瑞吉度假酒店共有 150 间装饰豪华的客房及 12 套宽敞的别墅与套房，其中还包括总统套房。酒店将西藏的丰富文化及异域自然风情完美地融合在一起，从设计之初便考虑到了诸多可持续发展方案，包括太阳能电板、酒店餐厅选用来自本地的农产品与草本并设有地下水循环系统。内部装饰精美的酒店客房配备了设计独特的家具、宽敞的大理石浴室、奢华床单以及诸如高速网络连接与等离子电视等设施。

瑞吉还将在拉萨瑞吉度假酒店首次推出最新的水疗品牌 Iridium Spa。旨在成为世界最高奢华级别水疗品牌的 Iridium Spa 能够令宾客彻底放松身心，重新焕发活力，其专为个人定制的水疗体验则包括传统西藏草本护理以及使用本地滋养草本与植物（例如柏树叶及杜鹃花叶）的香薰理疗服务。Iridium Spa 水疗中心融合了五星奢华体验以及神秘的西藏精神疗法，这里的恒温泳池富丽堂皇，闪耀着金色光芒。另外，宾客还能够在

沉思园中锻炼瑜伽、进行普拉提训练或者静思冥想。

拉萨瑞吉度假酒店的餐厅装饰极具文化气息，而这里的多种美食体验也会让人胃口大开。酒店的招牌餐厅秀(Social)拥有各类国际美食，全天候为您提供便利的餐饮服务；而斯自康则是第一家真正的藏式餐厅，这里环境优雅，令人倍感舒适，纯正的西藏美食与尼泊尔佳肴定会令你垂涎不已。宴庭（Yan Ting）共有六间私人包厢，宾客们能够在这里品尝各式各样的粤菜与川菜佳肴。

作为西藏首家奢华酒店，拉萨瑞吉度假酒店荣耀推出本地区第一家 Decanter by Haut-Brisson 酒吧，这里藏有 140 多种各类美酒以及陈年雪茄。品茗轩典藏了多种当地与进口名茶，而拥有 10 至 25 年茶龄的陈年好茶则更是令人陶醉。西藏也是第一次能够拥有这样的别致服务。

Lhasa is regarded by ancient Tibetan as "a land of the gods", and it is the Tibetan spiritual and political center. Located in the old Barkhor Street in Lhasa, St. Regis Resort is conveniently located, surrounded by all kinds of cultural landmarks, where its guests can explore the ultimate charm of Lhasa among these legendary buildings. This stunning resort hotel is surrounded by beautiful natural environment and stirring mountain views of Lhasa, and it is committed to presenting higher quality of service standard and luxury enjoyment to Tibet. The magnificent resort hotel provides guests with the characteristic St. Regis butler service, three World-class restaurants integrating international food, one famous teahouse, one bar and one Iridium Spa in an area of 11700 square feet. The swimming pool here is beautifully decorated and exudes golden lights. In addition, guests can also practice yoga, Pilates or meditate in the meditation garden.

St. Regis Lhasa Resort totally has 150 luxuriously decorated rooms and 12 spacious villas and suites also including the Presidential Suite. Considered with many sustainable development programs at the very beginning, the design for this Hotel results in a perfect harmony between rich culture and exotic natural style of Tibet, for examples, the hotel has solar panels, offers the food available in restaurant of the hotel made from local farm products and herbs, and operates a groundwater circulation system. The hotel rooms with beautiful interior decoration are equipped with unique designed furniture, spacious marble bathrooms, luxurious bed linen, as well as facilities such as high speed internet access and plasma TVs, etc.

St. Regis will also initially launch Iridium Spa, the latest spa brand at

the St. Regis Lhasa Resort. For the purpose to become the world's highest luxury level spa brand, Iridium Spa can help guests to completely relax body and mind, re-energize vitality, whose spa experience designed for individual customization includes traditional Tibet herbal care as well as aromatherapy physical therapy service using local nourishing herbals and plants (such as cypress leaves and azalea leaves). Iridium Spa center combines five-star luxury experience and mysterious Tibetan spiritual therapy, where swimming pool at constant temperature is magnificent, shining with golden lights. In addition, guests can also practice yoga, Pilates or meditate in the meditation garden.

The decoration of St. Regis Lhasa Resort's restaurants is extremely full of cultural atmosphere, where a variety of culinary experiences also make a good appetite. Social, the hotel's classic restaurant, has all kinds of international cuisine, providing you with convenient food and beverage services around the clock; while Sizikang is the first pure Tibetan-style restaurant with elegant environment, where people feel more comfortable, and pure Tibetan food and Nepalese cuisine will make you salivate. Yan Ting has totally six private dining rooms, where guests can enjoy a wide range of Cantonese and Sichuan cuisine.

As Tibet's first luxury hotel, St. Regis Lhasa Resort Hotel has gloriously launched the first Decanter by Haut-Brisson bar in local collecting more than 140 kinds of wine as well as vintage cigars. Teahouse collects a variety of local and imported tea, where the good tea collected for 10-25 years is even more intoxicating. Tibet is also the first time to have such a unique service.

花间堂·周庄季香院

ZHOUZHUANG BLOSSOM HILL BOUTIQUE HOTEL

设计 / Thomas Dariel

设计单位：Dariel Studio
项目经理：陈一凡
项目地点：周庄
建筑面积：2500 m²

花间堂季香院酒店位于江南水乡——周庄，粉墙黛瓦，厅堂陪弄，临河的蠡窗，入水的台阶，在这里，千年的历史也隐在江南迷迷茫茫的烟雨中，其温婉绰约的神韵随着碧波在不经意间一波一波地荡漾开来。离上海仅1.5小时的车程使其成为上海背后避世休闲的绝好去处。

这个精品酒店项目是由三幢明清风格的老建筑改造而成，相传这三幢老建筑曾分别属于戴氏一家兄弟三人，各自在这宁静的小镇经营其营生。时至今日，经过岁月的年轮无情地碾过，在改造之前这三幢独栋建筑分别被用做博物馆、茶室、客栈，并有一部分已废弃。Dariel Studio 非常小心地对这些优秀的古建筑进行修复并将其合并改建成拥有20套客房的精品酒店，同时希望可以保留建筑最原始的空间结构以及其历史传承。

昔日的戴宅被分为东、西、中三宅，三宅独立而建，却又紧贴相连，成为一个整体，格局迥异，各具特色。Dariel Studio 用了近半年的时间对其进行修复改建，包括地面高低的统一，主梁的加固，门窗的修复和重建，结构的重新划分等。

根据客户的要求，此精品酒店将与周庄如画风景和历史相结合，体现这古镇古往今来一直未变的恬静而优雅的生活，保留并延续当地的历史文化。

穿越季节的感官之旅

为了契合周庄这一古镇的历史感及中国文化元素，热爱中国文化的设计师Thomas Dariel 将这个精品酒店的设计主题设定为"穿越季节的感官之旅"，其灵感来自于中国传统的二十四节气。在中国，依照古时农耕的作息习惯和自然规律，人们根据太阳在黄道上的位置把一年平分为二十四个节侯，如立春、谷雨、小雪等。

首先，在酒店各房间的布局上，根据日照上升降落的分布规律，至南向北地将春夏秋冬依次在各排房间进行演绎。从浅浅的大地色，到跳跃的橘色，过渡到深沉的紫色，演绎了四季的不同个性。采用四季迥异的花卉来命名不同的客房名，芷樱、碧荷、丹桂、墨兰……并且运用不同的软装和灯饰来诠释，仿佛终于找到了隐世的桃花源，置身于四季的国度，久久不愿离去。

其次，设计师选取了几个重要的节气进行分别的表现，使整个酒店的空间立体分布具有季节性的标志。

春分，春暖花开，岸柳青青。用这一个节气来表达带领来访者进入一种新鲜的入住体验的接待处，那是最恰当不过了。春分，昼夜平分之意，也能恰如其分的代表接待处这一贯通里外的作用。

夏至和冬至，色彩的强烈对比和西式吧台与中式家具的搭配，表现出中西餐厅美食文化的激情碰撞。芒种，麦子丰收，酒醇得正是时候，也如同红酒一般令人珍惜，将红酒吧用芒种来形容是再贴切不过的了。

惊蛰，用于表现适合冥想的地方——阅读室，在这里聆听自己的内心，幡然醒悟更多哲理。悬于走廊的笼状灯笼、淡金的配色、舒适的沙发无一不使你希望在这个电闪雷鸣、虫儿苏醒的节气到来时，窝在这里品着香茗，读本好书。白露和小暑，分别代表水吧和茶室。白露象征了水的洁净与润泽；小暑正显示了茶所需要的温度。在这里，你可以体验不同季节带来的非凡感官体验，感受令人激动的时光之旅。

历史文化的保留及传承

为了更好地保护当地文化和传统建筑，在进行修复改建时，小到一砖一瓦一石子都被编号保留起来，并且修旧如旧，重现新生。那些实在毁坏严重的，Dariel Studio 采用相同形状花纹进行重新制作以符合明朝风格。在庭院门楼上秀才陶惟坻的题字“花萼联辉”，寓意戴氏兄弟连心，必能兴旺家业；堂楼之间天井相连，雕梁画栋，颇有气魄。正厅对面为三

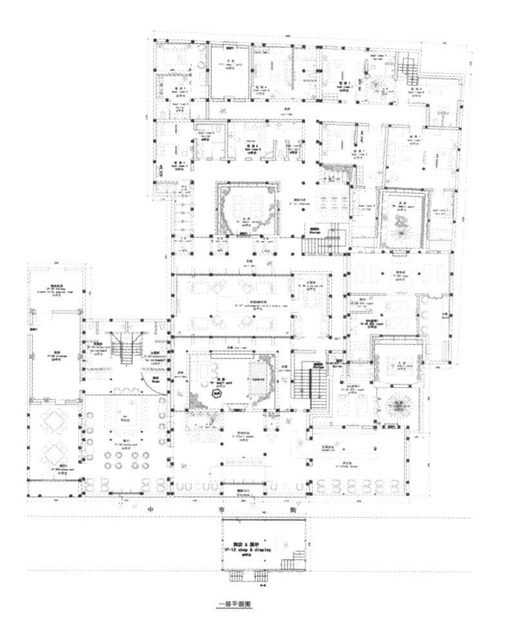

一层平面图

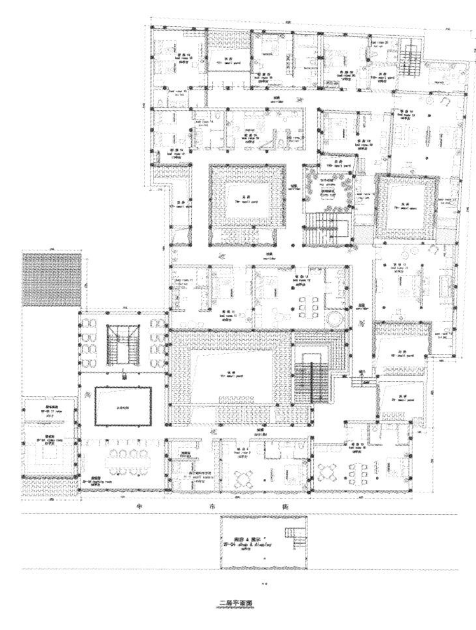

二层平面图

进重檐风火墙门楼，题额刻有“剡溪遗泽”，意为怀念故乡情深；房间里也随处可见留下的弯曲悬梁和雕花题刻。

除了保留原有的能被保留下的，更多的仿古装饰被装点进了这个空间，以切合整体的环境。灵感来源于中式手提食盒的各类柜子，支柱为竹子形状的中式大床，古式的门把，卫浴间门上的雕花铜片，镶在墙上的民族项链，各式的中国瓷器花瓶，亮色的交椅，墙上用各色毛笔笔刷高低错落排列作为装饰……

<u>中式传统优雅和西式当代奢侈的完美结合</u>

除了保留和恢复其中式的特点，来自法国喜爱将中法文化结合在一起的 Thomas Dariel 也将中西融合在这个室内设计中表现得淋漓尽致。代表惊蛰节气的阅读室里，配以一个西式壁炉和小型钢琴，不但在中式的氛围中更添一份温暖，更让人有种置身于法国文艺复兴时期的感觉，如此取长补短的结合真是恰到好处。中西餐厅的强烈色彩对比，巨大的悬式吊灯，以及那引人注目的法式花饰瓷砖砌成的吧台让人在一片传统中找到新鲜亮眼之处。更特别的是，装点墙面的各种画像也充分向中国丰富的手工艺品致敬。同时，“水中的墨滴”摄影作品表现出一种结合中国传统水墨书法以及当代的诗歌的感觉。各种中式装饰以及西式装饰的互相混搭营造出现代与古典的别样完美的结合。

<u>每个公共空间都是独立的个体</u>

在不同的公共空间，皆会产生不同的体验之旅。作为一个精品酒店，当然不同于传统的5星级酒店，在这里，你可以全身心的放松并且如同在家般自在地度假。这里有红酒吧、阅读室、影音观摩室、水吧、水疗、瑜伽室等活动中心满足休闲享受的需求。每个空间都将带您通往享受之旅。

充分利用空间与光线、展现本土文化与传统、精心选材并钟情于艺术与手工艺品，Dariel Studio 通过这种种手段使朴实与典雅交相辉映。这一空间的主要设计概念是永恒之时、静谧之美与精妙奢华。我们笃信舒适并非在于区区的富贵，而是应当体现在精致与珍奇之中。

在这幢典型的19世纪建筑中，我们将古老中国的特色与现代摩登家具交织在了一起，伴着当地特色的手工艺品，使游客沉浸于这座城市的灵魂与个性之中，并在现代的视觉效果与精致优雅的氛围下恢复活力。

Zhouzhuang Blossom Hill Boutique Hotel is located in Zhouzhuang, the water town of Southern China positioned 1.5 hours away from Shanghai. This ancient town, known as the "Venice of the Orient," has a history of more than 900 years and still retains the style and pattern of its ancient village. It is noted for its profound cultural background, the well-preserved ancient residential houses, the elegant views of the waterside and its colorful local traditions and customs. This hotel is a renovation of three old separate Ming Dynasty-style buildings. During the 19th century the three sons of Dai built the buildings that, at the time, served as living places, as well as a space to conduct business. After many years, these buildings were divided into four parts before the renovation: a Museum, a Tea House, a Guesthouse and an abandoned structure.. Dariel Studio has reconstructed and transformed them into a 20-suite heritage hotel. The renovation aims to restore the initial spatial unity of the building, while also to preserve the architectural heritage. These four buildings were independently built but closely connected, which is why the structure and style are similar. Dariel Studio spent nearly six months on repairing and altering the buildings, which include the adjustment of the ground level, the reinforcement of the main beams, the preservation of the old wooden window frames and the layout of space. Dariel Studio was asked by Blossom Hill to create a boutique hotel that embodies both the picturesque scenery and history of Zhouzhuang. These qualities combined portray the same elegant leisure-lifestyle experience that has endured through the ages in this ancient town, in addition to exhibiting its identity and history.

One sensory travel through the seasons

In order to apply the feeling of history and Chinese culture, designer Thomas Dariel created the "one sensory travel through the seasons" concept for this boutique hotel. The inspiration comes from the 24 seasons from the Chinese traditional solar calendar. In ancient China, the solar year is divided into 24 terms according to Chinese traditional farming habits, with each term corresponding to the Sun's particular position,. First, the rooms are divided into four seasonal areas and are displayed according to their sun exposure and natural light intensity during the day. Thus, these rooms symbolize the seasons Spring, Summer, Autumn and Winter. From light yellow or bright orange, to deep purple, the subtle choice of colors represents the atmosphere and characteristics of each season. Each room is named after a flower based on a different season, such as Blossom, Lotus, Sweet Olive and Cymbidium. Dariel Studio chose several important solar terms to design the hotel's entire spatial distribution, where each room is defined by a poetic seasonal symbolism: Vernal Equinox, Solstice of Summer and Winter, Wheat Heads, Awakening of Insects, White Dew, Little Heat…

"Vernal Equinox", for instance, welcomes visitors to enjoy a new experience. "Equinox", meaning an equal time between day and night, is emulated by a perfect balance between the exterior and the interior of the ho-

茉香

tel. -"Solstice of summer and winter" is defined by a restaurant symbolizing the duality of the two seasons. Furthermore, the representation of combining the East and the West is reflected by a Western-style bar accompanied by Chinese-style furniture.

The season of "wheat heads" is portrayed by the architectural design of a wine cellar. The intended experience of this room is represented by the time of harvesting plentiful amount of wheat. The maturing process of wheat heads and wine are similar, for both need time in order to develop flavor. - "Awakening of insects", a time when hibernating insects awaken, is a place of contemplation. The room awakens people's spirituality, mirroring the startling insects, with different attributes that define the space: the insect cage-style lamp, the golden color and the comfortable sofas intend to make the room a place of constant change.

"White dew and little heat" are periods differentiated by opposing temperatures. "White dew", a time of the year when it becomes rainy and cloudy, is symbolized by a water bar. The term "little heat", a time in July when the season's heat begins, is signified by a tea house. The warmth of tea and the refreshing water compliment the hot and cool seasons of the Chinese Lunar Calendar.

The preservation of cultural and architectural heritage

To protect the building's heritage and local culture, every old decoration, window frame, stone, and tile were carefully collected and noted before renovation. All of these materials were reused to preserve the aspects of cultural tradition. For the ones that could not be reused, Dariel Studio reproduced them with the same pattern to retain the Ming-dynasty style. An inscription, "hua e lian hui", by a scholar in ancient China was kept on the top of the archway. ,"hua e lian hui" translates to "the three brothers reaching success if they work together with the same ambition." An inscription"shan xi yi ze" on top of another archway means "the hometown is deeply missed." The main structure of this old buidling was also kept; patios connect the front halls and back halls, and Chinese pattern-carved beams and pillars reveal the building's historicity. Dariel Studio not only preserves the heritage of the area, but also emphasizes historical aspects of the building by artistically incoporating new Ancient-imitated decorations in the design. The shape of the cabinet and bathtub is conceptualized into the square shape of a Chinese ancient food-box; a Chinese king-size bed is kept company by a bamboo-shaped prop; there are Chinese metal handles on the doors and engraved copper on the door of bathroom; a special national necklace is featured on the wall; different Chinese vases are ubiquitous in the rooms and public areas; new brightly-painted Chinese chairs are present; and vibrant calligraphic brushes are hung on the wall for decoration.

Combination of Chinese heritage and Western Contemporary Luxury

Thomas Dariel, who likes to mix Chinese and French culture together, plays with the combination of Chinese traditional culture and Western modern luxury perfectly in this project. The library contains a fireplace and a baby-grand piano that not only make one feel warm, but that also portrays a calm and comfortable atmosphere similar to traditional Parisian salons—a space where one would spend all day reading or thinking. In the restaurant that identifies "solstice of summer and winter", the apparent color contrast of the repainted Chinese Ming-Style chairs set beside the bar with typical French pattern ceramic indicates the opposite seasons. In particular, the photographs created intentionally for the hotel, are a tribute to the richness of Chinese craftmenships and artpieces. As well, the "Ink in Water" photographs illustrate a blend of traditional Chinese ink calligraphy and poetry while featuring a contemporary touch.

One experience for each public room

Through strategic use of space and light, the design base on local cultures and traditions, choice of materials and a special love of art and crafts, Dariel Studio wants to combine modesty with great elegance. The main design concepts of space are synonyms to timelessness, calm beauty and subtle luxury. We believe that comfort does not lie in trivial affluence but in delicacy and rareness. One will go through a totally different experience when he or she stays in each public room. Rather than a 5-star hotel, a person could relax and spend time in a boutique hotel, all while wine tasting in the wine cellar; reading in the library; watching movie in the cinema room; tea tasting after attending a healthy spa or yoga. Each room will lead the guest on an enjoyable trip through a day by immersing travelers into the soul of the city.

I DO 汽车旅馆
I DO MOTEL

设计　程绍正摇

设计单位：真工设计工程股份有限公司
Z-WORK Design Associate
参与设计：张嘉伦、陆居赋、彭佩玲、
陈美惠、杨元仁
项目地点：台湾桃园
建筑面积：3220 m^2
主要材料：石材、木材、景观石、白砂、
刷石子、磨石子、铁件、玻璃
摄 影 师：汤Marc

台湾常见许多标榜Design Hotel甚至是五星级饭店式豪华气派的Motel汽车旅馆，但是像I DO顶级会馆做得这般彻底的则是头一个。台湾空间设计界大师级人物程绍正韬先生出马操刀、一手包办，从概念、视觉、室内设计到建筑结构、景观建造与空间氛围营造，全方位打造舒适怡人又令人印象深刻的Resort精品旅馆。

舒放身心的休旅文化殿堂

位于台湾桃园经国路上的I DO顶级会馆完全打破汽车旅馆华丽奢靡的情欲刺激，舍弃了缤纷的色彩所勾勒出极具情欲张力的赤裸感官印象，企图以“City Resort”概念出发，希望提供给客户不只是个下榻休息睡觉的地方，更是一个舒放身心和体验精致休旅文化的殿堂。

I DO顶级会馆的经营理念把重点放在人身上，希望能传递“美好、精致和值得信赖”的住宿体验给前来的房客，为了达到这个目标，在I DO 56个空间风格里，有最新的饭店设备，有最奢华的卫浴空间，也有最质朴原始的砖墙，基本上所有的设计都是以营造一个舒适、精致休闲生活为出发点，并不是抱持着漂亮就好，反正客人休息一、二个钟头就走了的华丽心态。

像家一般的舒适空间，能够较长时间的停留，是旅馆主人与设计师想出的创新概念。

为了让每一位住进来的房客感受到真正放松的休旅生活，设计师以“City Resort”为概念规划了56个空间型态，完全打破空间惯有的平衡，在偌大空间中，以浴室为主体，空间、家具在比例、尺寸上的放大，让人感觉新鲜、舒朗、开阔，创造了无限惊喜。

56变的多元化独立空间

设计就是以没有预期到的惊喜迎接你。

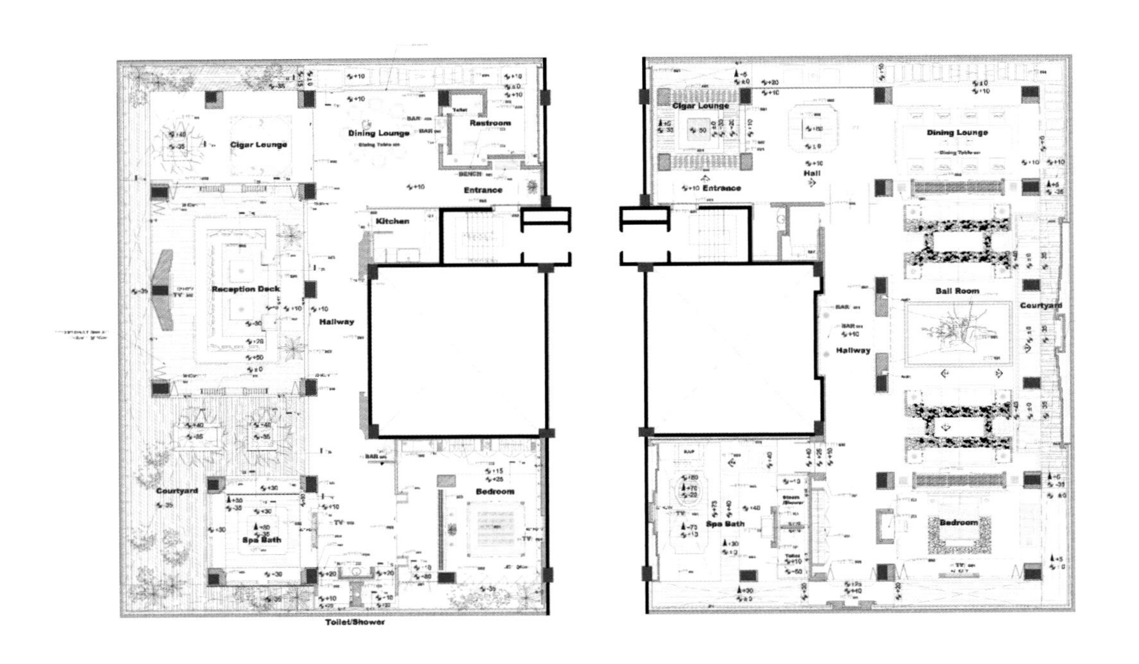
Cigar Lounge
Dining Lounge
Restroom
Entrance
Kitchen
Reception Deck
Hallway
Courtyard
Spa Bath
Bedroom
Toilet/Shower
Cigar Lounge
Dining Lounge
Hall
Entrance
Ball Room
Courtyard
Hallway
Spa Bath
Bedroom

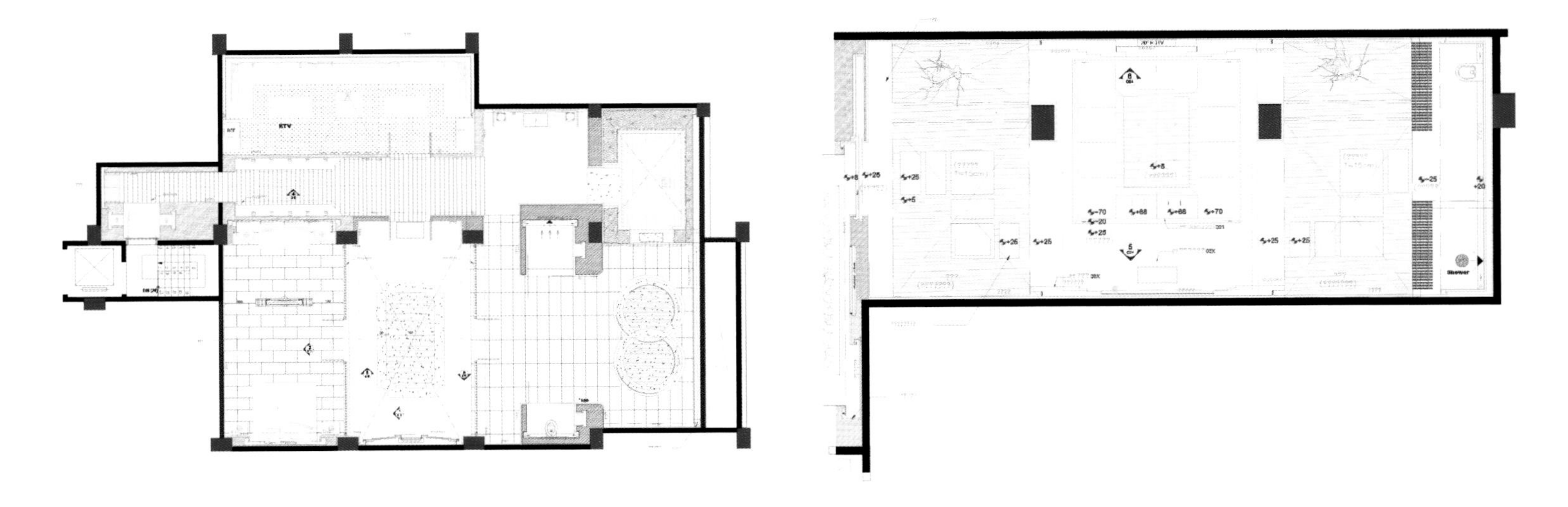
KTV
Shower

耗时一年多，耗资5亿台币兴建的I DO顶级会馆，有56间住房，设计师为她创造56种风情，56种在房间内旅行的方法。既见传统也见现代，既交融时尚又纳入简敛，既奢华贵气又自然质朴，融合各种趋向的矛盾性格，设计师运用各种材质做对比，无论是亚热带岛屿的风情屋或是巴里岛Villa的概念屋或是充满神秘隐晦氛围的时尚空间，没有因为工作量大就找可以以量制价的系统家具，每一间房，都有自己独特的空间内容与个性化感受，包括沙发、桌椅、橱柜等家具，更是全都由程绍正韬先生亲自设计精心打造，每个房间都是自成一格的艺术空间，都在挑战大家对于旅店的惯性期待。

所谓的旅店房间，其实更像是一间间独立、设备齐全的屋子，空间宽敞阔大，除独立客房、客厅、卫浴外，每间房必有完整成块的大理石拼合出光滑的超大浴池，还有瑰丽耀眼的马赛克磁砖点缀，不禁让人怀想起古罗马时代至高无上的澡堂文化；最迷人处是被四面包围的中庭、露天浴池，密封之中，却又流露著开放，就算长时间赖在屋内不想出门，也丝毫不显得拥挤压迫，更不会让人有被关起来的拘束感。这种彷佛置身别墅Villa中的静秘感，是四面八方而来的，不只是透过眼睛的观察，更掌握了空间内在隐微的变化，保留传统汽车旅馆的影子，透过精致的质材、细腻的手工艺，在冷硬材质交叠运用中表现简单、平静、温暖，巧妙组合出空间的开阔感，让总是被时间追着跑的烦躁，渐渐也慢了下来，如进入梦境般地安静。

在56种可供选择的不同房型中，最值得一提的是如Villa般的中庭屋。中庭屋的面积约40~60坪（1坪≈3.3 m²），它被设计为拥有户外中庭、露天温水游泳池、浴池、雨洒淋浴间和蒸气室、及独立的客厅和卧室。在这一系列的空间，设计师表现出对奢华的细节要求，从建筑形式、建材、物件与古董的收藏，示范了

对于东与西两方哲学的吞吐容纳，在极简框线中，穿插如自然岩石纹路的砖墙、石板，创造出迷漫的放松感，有点像在巴里岛般的开阔感，但却没有渡假的过度阳光，反而如私密之所的柔和，一种城市中的静秘之所，如此充满洁净氛围与丰富内涵的空间，同时满足了眼睛与心灵的渴求。更特别的是，在这寂静的天趣之间，原来一切可以变得更开放，例如洗手间、浴室，都可以设于室外而不必担心别人的眼睛，在这，你可以享有百分百绝对个人的清静，在可触的蓝天白云或在夜晚的星空下，一边泡在双人尺寸的超大独立浴缸与情人共洗奢华的鸳鸯浴，一边享受着自然天光与绿意洒落，这种将室外室内空间打成一片，营造出舒适而温暖的氛围，为的是一份城市人得来不易的私密和幸福感。

再深入每一个房间，千万别错过Room610房的另类体验，融合了阿拉伯、清真寺文化设计的Room610，纯以方正的几何块面、线条支架起这坦荡荡的偌大空间，步入房间的当儿，一种豁达平静的感觉袭人而来，教人俨如走进庙堂一般的宁静，然后得到净化。长廊中刻意地以烛光取代灯光，也让空间飘散着一股迷人的神秘气质，大量的白与极少的装饰让整个空间显得更为澄净，这样的空间教每一个走进其中的人，自自然然都成了个中主题。一如那一堵堵石墙上，你道是毫无装点的淡寡，而太阳的光来了，树木的枝桠、天上的蓝天白云便掩照在那素面之上，分分明明的，仪态亦随着光的明与灭而转化万千，这，便是丰盈。

City Resort也是一种空间艺术

不管是现实层面也好，文化意识层面也好，Motel汽车旅馆一直以来总予人一个通往情欲世界的印象，I DO顶级会馆的经营理念，却是优雅与放松的完美平衡。56种不同风格的住房，一种放松生活的态度，如果你想远离俗世喧扰，过一个不用载假面具交际应酬的周末假期，在这里可享受一段近乎奢侈的悠闲时光，不论房里跃动的是天马行空的想象，还是最私密不可告人的欲望隐喻，还是如豪宅般的贵族享受，I DO顶级会馆绝对让你在惊叹之余，久久舍不得离去！所以，邀请情人或朋友或亲人来渡过一个令大家惊喜的慢活假期吧！

In Taiwan, there are many motels which boast to be design hotels or which is as luxurious and stylish as five-star restaurants, but there is never a hotel so perfect like I DO super club house, which is exquisite in every aspect. I DO super club house was established under the supervision of Taiwan' s space design master Mr. Chengshao Zhengtao. From concept, visual effect, interior design, architectural construction, to landscape planning and space atmosphere, Mr. Chengshao Zhengtao helped create an all-around comfortable, pleasant and impressing resort hotel.

A relaxing and recreational cultural palace

I DO super club house which locates on Jingguo Road of Taoyuan, Taiwan has no erotic or extravagant flavor of ordinary motels, and abandons the erotic impression created by multiple bright colors. Built upon the City Resort concept, the club house not only provides guests a sleeping place, but also a tasteful palace where the guests can relax themselves and experience the exquisite recreational culture.

With its operation philosophy centered on people, I DO super club house hopes to deliver a beautiful, exquisite and trustful lodging experience to guests. To achieve this, the club house provides 56 living styles featured with the latest restaurant equipment, the most luxurious bathroom, or simple and original brick walls. All of these styles aim to create a comfortable, exquisite, and recreational living experience, and not just superficially beautiful space merely for a couple of hours' rest.

According to the innovative concept forwarded by the owner and designer, the club house should be as comfortable as a home, and pleasant for a long period of stay. In order for the guests to totally relax and enjoy the recreational living, the designer plans 56 types of rooms based on the City Resort concept, breaks the customary space balance, and makes bathroom as the main part of the large space. The enlarged size of space and furniture will make guests feel fresh, bright and open, bringing infinite surprise to them.

56 different types of independent space.

Design is unexpected surprise.

I DO super club house cost 500 million Taiwan dollars and took more than a year to be completed. The club house has 56 rooms with the 56 different styles. The rooms are both traditional and modern, fashionable and reserved, luxurious and simple, integrating various kinds of contradictory natures. The designer uses many different kinds of materials to implement the contrast. No matter the fascinating subtropical island room, Bali villa concept room, or mysterious and obscure fashion room, every room has its own special content and personalized feeling, and no room is stuffed with cheap, mass production outcomes. All rooms' furniture, including their sofa, tables, chairs and closets, are designed by Mr. Chengshao Zhengtao. Every room is artistic in its own way, challenging people's customary expectations on hotels.

The hotel rooms are like independent and fully equipped homes with abundant space. Except for the independent guest room, living room and bathroom, every other room has a super large and smooth bathing pool made of complete marble blocks and ornamented by brilliant mosaic tiles. The bathing pools will remind people of the ancient Rome's supreme bathroom culture. The most attracting is the outdoor bathing pool in the courtyard, expressing openness in the enclosure. Even if you stay in the room for a long time, you will not feel constraint. You will just feel so quiet as if you are living in a villa. This tranquility feeling comes from all directions and not just from what you can see. The delicate changes in the space which are inherited from traditional motels, the exquisite materials and handicraft, and the overlapping of cold and hard materials all contribute to the openness of space and the simple, tranquil and warm feelings. In this kind of atmosphere, you will just put down the agitation of hurrying, and step into a quiet realm like being in a dream.

The most distinguished of the 56 types of rooms is the courtyard room which feels like a villa. It takes about 40 to 60 square meters, and has an outdoor courtyard, warm water swimming pool, bathing pool, shower room and steaming room, and independent living room and bedroom. In the series of space, the designer expresses luxury in every detail from the architectural style, materials, objects and antiques, demonstrating both eastern and western philosophies. In the extremely simple frames and lines, the brick walls and stone tiles with natural rock patterns create relaxation feelings, like the openness in the Bali Island without excessive sunshine while at the same time the softness feelings of a private place. It is a quiet, clean and meaningful place in a city, satisfying the desires of both eyes and the mind. More specially are the outdoor bathroom and bathing pool. Here, you can enjoy the absolute quietness without worrying being seen by others. Under the blue sky in the day or at the starry night, have a luxurious bathing with your lover in the super large double-size bathing pool, and enjoy the natural light and green trees. The combination of outdoor and indoor space helps create the comfortable and warm atmosphere, in which the city people can enjoy the unusual privacy and happiness.

Moreover, never miss the special experience in room 610 among others. The room is designed based on the Arabic and mosque culture, and uses geometrically rectangular patterns and lines as the main elements of the large empty space. Walking into the room, you will feel open and calm as if you are in a temple and being purified. Candles light the long corridor instead of electric lights, which bestows a kind of charming mysterious atmosphere upon the space. Large amount of whiteness and very few decorations make the whole space feel clear and pure, making everyone in the room to be its theme. You may think that the stone walls that have no decoration are simply plain and dull, but when there is the sunshine, the blue sky, white clouds and branches of trees reflect on the walls, with their shapes changing and transforming along with the brightness and dimness, you will know what richness is.

City Resort is also a kind of space art.

No matter in reality or in cultural consciousness, motels always give an erotic feeling. Unlike them, I Do super club house creates the perfect balance between grace and relaxation. The 56 styles of rooms all symbolize an easy and relaxed living. If you want to leave the daily noise and spend a quiet weekend without social courtesies, you can come here and enjoy a period of luxurious leisure time. You can have unrestrained imagination in your room, satisfy your most private desires, or enjoy the nobility of living in a villa. Once you have lived in the amazing I DO club house, you will sure be unwilling to leave. So, invite your family, lover, or friends to spend a surprising and slow vacation here.

上海安达仕酒店

SHANGHAI ANDAZ HOTEL

设计　Takashi Sugimoto

建筑设计：Kohn Pedersen
Fox Associates
室内设计：Super Potato
项目地点：上海嵩山路88号
建筑面积：47144 m²
摄 影 师：Marc Gerritsen、申强

上海安达仕酒店坐落在上海市中心最著名的娱乐、购物和商业中心——上海新天地。整栋建筑由世界顶尖设计事务所、总部位于纽约的Kohn Pedersen Fox Associates (KPF) 设计。充满创造力的设计，多重方案相融合，与当地环境完美衔接，上海安达仕酒店为现代城市高楼设计树立了新的标杆。KPF首席设计师Joshua Chaiken先生如此形容这个项目："最出乎意料的挑战在于要设计一个能与周边环境相融合的砖墙外立面。我们希望保持建筑的简洁风格，同时又能满足酒店运营多方面效能的需求，在设计时特别着重于石墙面的表现力和延伸感。"

酒店是令人瞩目的28层建筑，共有307间宽敞客房。上海安达仕酒店的客房安排多彩缤纷的惊喜特色。客房的窗户不但可以观赏上海新天地商区活力四射的景观，还能俯瞰整个城市。客房照明以气氛灯光为主题，创造令人印象深刻的效果：LED灯从天花板俯射整个房间，住客可以根据自己的喜好及心情随意转换室内灯光的颜色和气氛。浴室设计也别有机杼，墙面采用大巧不工的石凿设计，透明的盥洗池和半透明的浴缸亦可发出绚烂的彩光，地热系统为足底提供了舒适的体感。梳妆室与浴室分隔。

酒店的地下一层也使用了气氛灯光的概念作为整体设计的关键元素，15 m的半透明泳池与健身馆的镶板墙面均使用了内发光LED照明系统。

下行至地下二层，Optime Spa以"时光"的概念为灵感来源。无论客人是缺乏时间，还是可以享受充裕的时间，Optime Spa将根据客人的需求量身定做各种疗程，艺术而全面地疗养身心。这种超越时间的舒适体验将留驻在记忆里，即使回到喧哗烦嚣的快节奏都市依然无比受用。

置身于上海最著名的娱乐中心——

2nd floor

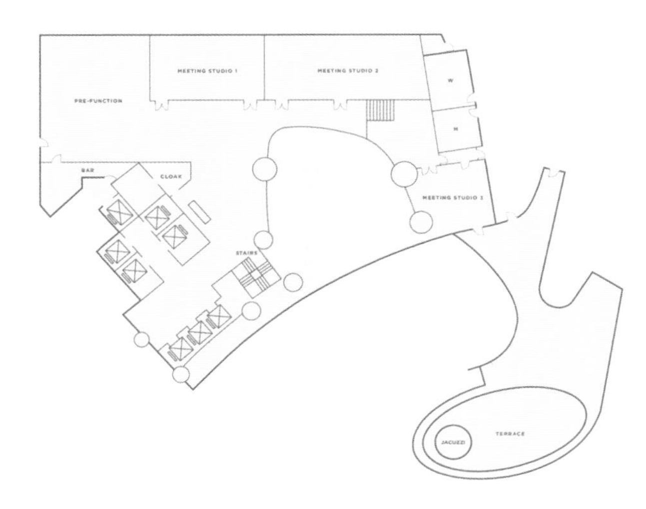

3rd floor

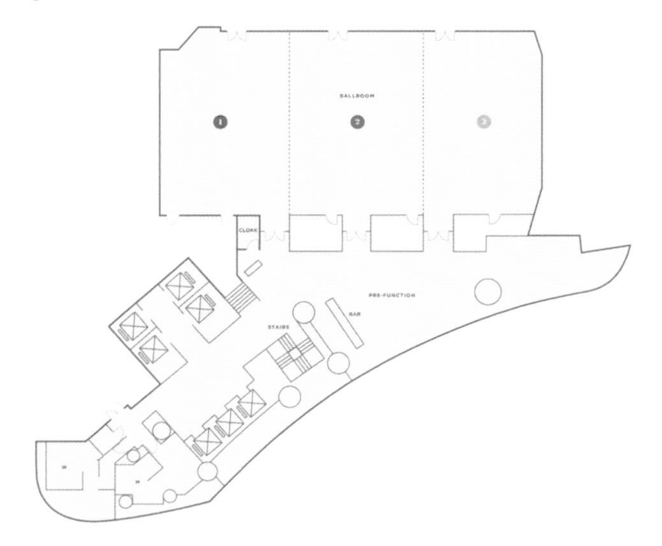

上海新天地商圈的心脏地带，上海安达仕酒店的餐厅和酒吧旨在展示充满活力的地方格调。餐厅秉持“从市场到餐桌”的概念，提供最新鲜、最合时令的本地食材。栖身大堂的安达仕酒廊深谙酒店的节奏——自由随性，全天候开放的酒廊让住客与访客们可以尽情畅叙，轻酌美酒或者品尝小吃。

占地四层的“海派”毗邻酒店主建筑楼——名字取“摩登上海精神”之意——集餐厅、酒吧、私人包厢于一身，是上海餐饮娱乐的新地标。位于一层的海派餐厅全天候开放，提供以沪菜为主的正宗中式菜肴和简单可口的法式美食，融合了上海风情的精致和法式小馆的随性浪漫，充分体现了酒店所在的地域特色。餐厅露台可以尽览上海新天地的繁华景象，为食客提供亲切舒适的用餐环境。

餐厅中餐区域设有开放式厨房，如剧场般向食客展示制作过程，这里有新鲜面食和粥品制作展示区和传统烤肉展示区。沪菜菜单上还有诸多经典菜式，如上海香糟扣肉、小笼包、龙井水晶虾仁等。另一方面，法式菜单提供各款经典的法国菜，如尼古斯沙拉、法式洋葱汤和冰镇海鲜拼盘等特色高卢开胃菜；主菜包括法式红酒鸡、顶级西冷牛排、布吉尼翁烩牛肉和油焖鸭腿；此外还有经典永恒的甜品，如杏子塔和柠檬塔。餐厅选用的高脚木桌、新鲜出品的食物、鲜活刺激的调味品，无不展现出独特的现代贸易集市风情。

海派餐厅在夹层设有4间私人包厢，每间可容纳6~16位客人。在二层和三层是海派酒吧。酒吧设计成私人家庭派对风格，让客人感到时刻包围在欢声笑语的房子里。屹立的巨型拉丝卵形钢塑将酒吧分成两层，均接受预约举办私人派对。

The Shanghai Andaz Hotel locates at the Shanghai Xintiandi which is most famous entertainment, shopping and commercial center of Shanghai. The whole building is designed by world top design office - Kohn Pedersen Fox Associates (KPF), with headquarter in New York. The design with full creative ability, the mixture of multiple schemes, and perfect combination with local environment, the Shanghai Andaz Hotel establishes a new model for modern city high-building design. The KPF chief designer, Mr. Joshua Chaiken, described such project, "The challenge which exceeds my expectations is to design a external facade of brick wall which can blend with the surrounding environment. We hope to keep the succinct style of such building, and also meet various efficiency demands of hotel operation, that we emphasized the expressive force and extension feeling of stone wall surface during design.

The newly constructed 28-storey hotel features 307 spacious guest rooms. The hotel guest rooms arrange colorful, riotous and pleasant features. The windows of guest rooms can

watch the vigor landscape of Shanghai Xintiandi commercial district, and can also overview the whole city. The theme of lighting of guest rooms is atmosphere lighting, to create impressive effects: the LED lamps light whole room from ceiling, the guests can randomly switch the inside lighting color and atmosphere according to their preference and emotion. The design of bathroom is original, the wall surface adopts the unskillful stone-cutting design; the transparent basin and semi-transparent bathtub can also send out the gorgeous and colorful lights; and, the floor heating system provides the comfortable feeling to foot bottoms. The dressing room is separated from the bathroom.

The 1st basement floor also adopts the atmosphere lighting concept as the key element of overall design; the 15m semi-transparent swimming pool and paneling wall-surface of health-building room adopt the internal-shining LED lighting system. On the 2nd basement floor, the Optime Spa utilizes the "Time" concept as the inspiration source. Even if the guests do not have much time or have enough time, the Optime Spa can establish various treatment courses according to their demands, to artfully and completely recuperate their minds and bodies. Such comfortable experience beyond time will stay in your memory, and will be still useful even you go back to noisy and rapid cities.

It locates at the Shanghai most famous entertainment center——heart of Shanghai Xintiandi commercial circle; the restaurants and bars show the lifeful local pattern. The restaurants adhere the concept "from market to table", to provide the most fresh, seasonable local ingredients. The Andaz Lounge in the lobby knows the hotel rhythm——the free, random and 24h-open Lounge can let the lodgers and visitors talk freely, drink a little good wine or taste the snacks.

The "Haipai" occupies four floor and is close to the hotel main building, the name is from meaning of "modern Shanghai spirit"; it integrates the restaurant, bar and private box, and is the new landmark of Shanghai restaurant and entertainment. The Haipai restaurant on 1st floor operates 24h per day, provides the true Chinese-style dishes with the Shanghai food as main theme, and simple & delicious French-style foods, combines the fineness of Shanghai flavor and romantic of French little restaurants, and fully reflects the local features of hotel location. The restaurant terrace can view the whole prosperous scene of Shanghai Xintiandi, and provide the warm and comfortable dinning environment for eaters.

The Chinese-style food area has the open-type kitchen, to show the preparation process to the eaters like the theater; there are the fresh wheaten food and gruel preparation exhibition areas and the traditional toasting exhibition. The Shanghai-style menu has many classic dishes, such as the Shanghai vinasse & port, small steamed bun, Longjing Sauteed Crystal Shrimp. On the other hand, the French-style menu provides various classic French dishes, such as Nicoise salad, French onion soup, cold seafood platter and other distinguished French appetizers; the main dishes include the

French-style red-wine chicken, top-class sirloin steak, Bourguignon braised beef and oil-braised duck-leg; and, there are classic sweetmeats, such as the apricot tart and lemon tart. The restaurant uses the high-leg wooden tables, fresh foods, and delicious condiments which show the unique flavor of modern commercial market.

The Haipai restaurant has four private boxes, each can hold 6~16 guests. The 2nd and 3rd floors are the Haipai bar. The bar is designed into the private family party style, to let the guests feel staying in a house with cheers and laughter. The erected large wire-drawn oval-shape steel model separates the bar into two layers. The bar accepts the appointment to hold private parties.

厦门海港英迪格酒店

HOTEL INDIGO XIAMEN HARBOUR

设计　Joseph Pang

设计单位：Joseph Pang
Design Consultants
项目地点：厦门
建筑面积：31886.96 m²

厦门海港英迪格酒店占据了厦门最具都市气质及人文底蕴的鹭江道CBD的核心位置，更与鼓浪屿隔海相望，将鼓浪屿的景致和一线无敌海景独揽。无论是酒店的客房还是多功能公共空间，设计上充分融合了厦门现代城市活力和传统历史人文韵味。

厦门海港英迪格酒店拥有128间客房，包括8间套房，每一间都经过精心设计。作为国际精品酒店品牌，酒店客房全部选用实木地板、TOTO全智能电子座便器、智能化窗帘系统等精装修高端配备。此外，每一家英迪格酒店在设计上都独具个性，融入酒店所在城市及周边地区的文化与历史特色。酒店每一间客房的设计都独具匠心，如宽敞的浴室空间和步入式淋浴体验、体现邻里人文装饰的墙面、趣致的画板、迷你吧和独特设

计的茶具等，都将为宾客提供生机蓬勃、魅力十足并且带着浓郁厦门邻里文化风情的旅居体验。

“江畔”位于酒店四层，贯穿上下两层，包括餐厅、江畔酒廊、图书室、上网空间等多个独立的活力空间，打破了传统酒店设施的界限，是宾客休闲与会客的绝佳场所。在这里既可享用富有当地特色的美食；也可在江畔酒廊细细品味别有韵味的时尚鸡尾酒饮品；或者在图书室阅读一本好书、观赏一场经典电影；以及在配备了Apple电脑的上网空间里畅享网上冲浪的乐趣。江畔的设计，采取了各种色彩与式样的混搭，不同的空间营造出的别样氛围，满足宾客休闲小憩、社交或是工作的不同需求。

而位于酒店五楼的健身中心，配备了齐全的健身器材，宾客可以在此进行有氧运动放松身心。

厦门海港英迪格酒店拥有4间会议室，以满足会议与活动的需求。其中一间全海景会议室，不仅阳光充裕，更以全幅的落地玻璃，将鼓浪屿景致尽收眼底。酒店还有厦门独一无二的可以俯瞰鼓浪屿全景的大型户外露台，宽敞的空间设计将宜人的海风引入，鼓浪屿秀美的风景如同画卷慢慢展开，是鸡尾酒会或BBQ等私人娱乐的完美之选。

The Hotel Indigo Xiamen Harbour locates at the core position of Lujiang Road CBD with most city temperament and culture foundation, faces the Gulangyu across the sea, and solely encloses the Gulangyu scenery and unmatched seascape. The design of hotel rooms and multifunctional public areas fully mix the modern Xiamen vitality and traditional history and cultural lingering charms together. Boutique

The Hotel Indigo Xiamen Harbour has 128 rooms, including 8 suites; each room is well designed. As the international boutique hotel brand, all rooms of such hotel adopt the wooden floor, TOTO completely intelligent water-closet, intelligent curtain system and other high-class facilities for fine decoration. In addition, the design of each Indigo hotel has individual character, involves the culture and historic characteristics of the city and surrounding districts. The design of each room has great originality, such as the wide bathroom space and stepping-in shower experience, wall surfaces with neighbohood culture decoration, interesting drawing-boards, min-bars and distinctively-designed tea sets, etc; all of these will bring the guests the mighty and seductive residence experience with Xiamen strong neighborhood culture.

The "River Bank" locates on 4th floor of such hotel, occupies two floors, includes the restaurant, river-bank lounge, library, net playing space and other independent activity spaces, breaks the boundary of traditional hot facilities, and are the perfect locations for guest leisure and receiving visitors. Here, the guests can enjoy the local special delicacies, taste the vogue cocktails at river-bank wine corridor, or read a good book or watch a classic movie in the library, and enjoy the pleasure of surfing on the Internet in the net-playing space equipped the Apple computers. The river-bank design adopts the mixture of various colors and types, creates unique atmospheres for various spaces, to meet the guests' different demands on leisure, social intercourse or working. The fitness center on 5th floor of such hotel has complete body building apparatus, the guests can have aerobic exercise to relax themselves here.

Such hotel has 4 meeting rooms to meet various meeting and activity requirements. One whole-seascape meeting room has abundant sunshine, and also has a panoramic view of Gulangyu with the whole-width full-height glass. The hotel has a sole large-scale outside terrace in Xiamen to overview the whole view of Gulangyu; the capacious space design introduces the delightful sea wind, the graceful scenes of Gulangyu are slowly unfolded like the picture scroll; it is the perfect option for cocktail lounge or BBQ and other private entertainments.

璞麗酒店

THE PULI HOTEL AND SPA

设计　Johannes Hartfuss、Jaya Ibrahim

设计单位：LAYAN DESIGN GROUP、
JAYA & ASSOCIATES
照明设计：THE FLAMMING BEACON
项目地点：上海

璞麗酒店地处南京西路与延安中路之间，位于上海繁华的静安区核心。璞麗酒店共有26个楼层，包括193间豪华客房及36间顶级套房，客房面积从45 m²到130 m²。位于酒店大堂一侧长达32米的"长吧"串联起大堂、户外水景、会客书廊及酒窖。酒店三楼是健康会所，配备了最新设备的健身房以及桑拿、蒸汽房，冰雾冲淋房和按摩池；会所另有长25 m的无边恒温泳池，能俯瞰静安公园。

璞麗酒店的标志，来源于其独特的原创定位——凤凰栖梧桐（凤凰树），代表着璞麗酒店的专属与尊贵。凤凰是中国神话故事中最为珍稀的至尊形象，她只择梧桐而栖，也因此成为尊荣的象征。而璞麗酒店的命名，鉴于中国文化中最显帝王尊贵的玉石，也因其具有永恒、璀璨的特性与质感成为中国人文美学的典范，所以千百年来一直受到人们的尊崇与珍藏。从璀璨原石到精雕的美玉，玉石的价值胜过其它宝石，如同天然璞玉般的纯净表征正是璞麗酒店的内在潜质。

璞麗酒店的室内设计是由杰出的澳洲设计公司LAYAN DESIGN GROUP联合世界闻名的印尼JAYA & ASSOCIATES室内设计公司及屡获奖项的澳洲灯光设计公司THE FLAMMING BEACON共同打造

整个酒店的公共地区、客房及景观设计。

设计借由材料的运用，巧妙地将中国文化融合其中，使新、旧感受并列且同时呈现出东、西方文化交融的独特风格。

汇聚中国古老元素与现代工艺科技的璞麗酒店使用了通常被用在建筑外墙的上海灰砖作为内部装修建材之一，营造出建筑的特殊美感与功能，酒店大堂的特殊地砖是由北京紫禁城修复工程地面建材的同一厂家所提供，并耗费很长的时间专为酒店精心制作的。这是设计师善用如今少用的传统建材作为空间永恒典范的又一构思。在每间客房内部随处可见的龙麟纹木雕屏风与铸铜洗脸台再度验证了设计师刻意交织古今与中西于一体的设计巧思。

酒店内各处，都可体验到室内灯光设计切合空间主题的巧妙思维。为了在这忙碌的都市生活中带来自然和悠闲的环境，The Flaming Beacon打造了一系列静态和动态的投影，创造出奇妙的空间氛围。独特的灯光效果与室内设计配合得天衣无缝。许多特别的灯具，其灵感来自于中国传统灯具外形，融入现代的灯具设计，达到意想不到的效果。

The PuLi Hotel and Spa is located between Nanjing West Road and Yan-An Road, at the very heart of busy Jing-an District of Shanghai. There are 193 deluxe rooms and 36 top-level suites covering areas ranging from 45 to 130 square meters on the 26-storey building. The "Long Bar" extending 32 meters at one side of the lobby functions as the linkage between the lobby, the outdoor waterscape, the library for visitors receptions and the wine cellar. The third floor of the hotel functions as the healthcare club, which has been equipped with a fitness center with the updated equipment, a sauna, a steam room , a jacuzzi pool and a spa; there is also a 25-meter-long rim-free thermostatic swimming pool in the club, which overlook Jing-an Park.

The logo of the hotel is derived from its unique design orientation- a phoenix perching on a Chinese parasol tree (phoenix tree),which represents the exclusivity and dignity of the hotel. Being the most rare and supreme image in Chinese fairy tales, a

phoenix only perches on a phoenix tree, and therefore it has become the emblem of dignity. The name of the hotel means the beauty of an unpolished jade, which best indicates the dignity of emperors in Chinese culture and serve as the example of littérateur owing to its everlasting, resplendent qualities and textures, admired and collected by the people for thousands of years. A jade, whether unpolished or carefully carved, is more precious than other kinds of gem and it is the pure appearance of an unpolished jade that is the inner quality of the Puli Hotel and Spa.

The interior of the hotel is designed by LAYAN DESIGN GROUP, an Australian Designers, with the common effort of JAYA & ASSOCIATES, an Indonesian Interior Designers famous around the world, and THE FLAMMING BEACON, an Australian Lighting Designers repeatedly winning awards, which designed the pubic area, guestrooms and different sights.

Chinese culture is integrated into the design by skillful applications of materials, so that the hotel uniquely feels both modern and traditional, with east and west cultures juxtaposed

The hotel, to which both ancient elements from China and modern technology applied, is provided with Shanghai gray bricks as one of the materials for its interior decoration, which is generally used in exterior walls, so that a special architectural beauty and function are presented. The tiles on the floor of the lobby are also specially supplied and carefully produced for a long time by the same factory that provided floor materials for renovation of the Forbidden City in Beijing. This is another plan of designers to make an everlasting spacial example with traditional construction materials rarely used at present. The wooden screen with dragon scale pattern carved on it and the sink made from cast copper found everywhere in each guestroom once again prove the masterly plan of the designers to deliberately merge the ancient with the modern, the Chinese and the west.

The ingenious design of interior lighting compatible with the spacial subject can be experienced everywhere in the hotel. In order to provide natural and leisurely environment in the busy city, The Flaming Beacon has created a series of projections, dynamic or static, to make a magic spacial atmosphere. The unique effects from the lighting work wonderfully with the interior design. Many light fittings are inspired by the shapes of traditional Chinese light fittings, with modern design involved, to achieve unexpected effects.

项目地点:三亚
客　　户:三亚鹿回头旅游区开发有限公司
摄 影 师:申强

三亚半山半岛洲际度假酒店
INTERCONTINENTAL SANYA RESORT

三亚半山半岛洲际度假酒店地处三亚风光最宜人的景区之一——小东海。苍山环抱，碧水白沙，美丽的亲水景观与水上花园遍布整个酒店，为酒店营造出一种别具风情的清雅之气。酒店共有343套客房，其中包括24栋独立的海滨别墅，1间总统套房和12间行政套房。宾客可随心享用酒店内的5个泳池，8间餐厅及酒吧，更可在洲际水疗会所内享受身心舒展的美妙旅程。此外，酒店设有1间无柱式宴会厅及7间会议室，巧思设计旨在满足各类会议与聚会需要。

三亚半山半岛洲际度假酒店的设计独具创意，是演绎洲际酒店集团崭新的度假酒店设计理念的第一份作品，其先进的环保理念独树一帜。酒店的总体规划体现了洲际酒店集团以客人为中心的设计思路，充分考虑到不同客人的独特习惯、特质和偏好，将整间度假酒店划分为几大区域，让不同类型的客人都能在此找到属于自己的空间，无论是商务、行政、休闲还是家庭度假客人，都能在此获得更为便捷和愉悦的体验。

酒店的设计充分考虑了它所在的热带环境，尽可能对自然光线和自然通风加以利用，减少对电力照明和空调的使用。在公共区域、走廊和餐厅，每一处可以开窗的地方都被充分利用，引入自然通风。宴会厅拥有充足的自然采光，大部分会议室都配有通透又宽大的落地窗和独立花园。公共区域遍布水景元素和盆栽植物。从客房及公共区域望出去，视线所及的屋顶平台全部为种类繁多的热带园林植物所覆盖，帮助减少热负荷，作为建筑的一层天然屏障，起到降温作用。

在为商务旅客设计的“天阁楼”客房区域，所有房间（包括浴室）均可看到海景，点缀着热带植物的阳台和自然通风的走廊亦是其中亮点。安置于“天阁楼”屋顶上方的太阳能热水系统不仅保证了这一区域所需的热水提供，同时也为非客房区域提供了热水保障。该系统每天可以节约 300 m^3天然气消耗，每天减少二氧化碳排放 600 Kg，按照使用 70% 天数计算，每年可减少二氧化碳排放量 153 t，这将非常有利于环保和实施绿色酒店的措施。

客房阳台和开放式走廊环绕着适合家庭休闲旅客的“水景房”客房区域和适合商旅客人的“洲际俱乐部”区域。宽阔的庭院，自然海风有效降低了这一区域的温度。庭院中大面积的绿植更为其增添丝丝清凉，同时也吸引鸟类的光临，营造出一片充满自然生机的绿洲。同样，这两个区域内的所有客房也拥有无敌海景，阳台延伸入水景中庭上方，空气更为流通，度假气氛更加浓郁。

洲际俱乐部泳池创新地采用了盐水，而非传统的经过氯处理的水，减少了化学物品的运用，对客人来说是一种新奇又环保的体验。“灵感天地”的休闲区及艺术廊为客人提供了挖掘并探寻独特的当地文化和历史的机会。

The Intercontinental Sanya Resort locates one of the most delightful scenic spots in Sanya – Xiaodonghai. Being enclosed by mountains, with blue water and white sand, the beautiful water scenes and overwater gardens can be found everywhere in such hole, to create an elegant atmosphere with special feeling. The hotel has 343 rooms, including 24 independent seashore villas, one presidential suite and 12 executive suites. The guests can randomly enjoy the five swimming pools, eight dining rooms and bars in our hotel; and, they can even enjoy the splendid trip with complete relaxation in our intercontinental Spa club. In addition, the hotel has a column-less banquet hall and 7 meeting rooms; these artful designs focus on the demands of various meetings and gatherings.

The design of Intercontinental Sanya Resort has specific originality; such hotel is the first works to deduce the completely new resort design ideas of Intercontinental Hotels Group; and, its advanced environment-protection idea is unique. The hotel overall planning shows the with-the-guest-as-core design idea of Intercontinental Hotels Group, and completely considers about the unique custom, peculiarity and preference of various guests, divides the whole hotel into several zones, and lets various guests find their own spaces; the commercial, executive, leisure and family-vocation guests can achieve convenient and delighted experiences.

The hotel design fully considers about the tropical environment of its location, utilizes the natural light and ventilation as much as possible, and reduces the usage of electrical lighting and air-conditioning. In public

spaces, corridors and dining rooms, each place where the window can be opened is fully utilized to introduce the natural ventilation. The banquet hall has adequate natural lighting, and, most meeting rooms are equipped with the transparent, wide and large French windows and independent gardens. The waterscape elements and potted plants can be seen everywhere in the public areas. Watch from the guest rooms and public areas, the roof platforms in your sight are coved by various tropical garden plants, to reduce the heat load; they serve as a layer of natural protective screen, and have the temperature-lowing function.

In the "heaven garrets" which are designed for commercial guests, all rooms (including the bathrooms) can view the seascape; the terraces with tropical plants and corridors with natural ventilation are also the highlights. The solar-energy hot-water system installed above the "heaven garret" roofs can ensure the hot-water supply for such area, and also provides heat-water for other area outside guest-room area. Such system can save gas consumption 300m3/day, and reduce the CO2 emission 600 kg/day; calculating with 70% utilization days, the CO2 emission can be reduced 153 t/year, it is beneficial in environment protection and implementing the green-hotel measures.

The guest-room terraces and open-type corridors surround the "waterscape rooms" which are suitable for family leisure guests and the "intercontinental club" which is suitable for commercial-trip guests. The wide courtyards and natural sea wind efficiently reduce the temperature of such area. The large-area green plants in the courtyards add a little of cool, and also attract the bird coming, to create the oases with natural spirit. Also, all guest rooms in the two areas have unmatched seascape, the terraces extend to the above of waterscape, the air is more smooth, and the leisure atmosphere is stronger. The swimming pool of intercontinental club adopts the salting water, not the traditional chlorine-treated water; it reduces the chemical utilization, so it is a strange and environment-protecting experience for the guests. The leisure areas and art corridors with "inspiration" provide the chance to excavate and seek the unique local culture and history to the guests.

香港唯港荟酒店

HOTEL ICON

设计　林伟而

设计单位：思联建筑设计有限公司
参与设计：温静仪、Jane Arnett、邓韵婷、谈健铭、萧宝珊
客　　户：香港理工大学
项目地点：香港
建筑面积：470 000 m^2
设计时间：2011.03
摄 影 师：Nirut Benjabanpot、Josiah Leung

香港唯港荟酒店(hotel icon)位于九龙主要旅游区，是香港理工大学设立的教学研究酒店，这里拥有262间客房，是一个可以看的到维港海景的酒店。除了地理位置绝佳之外，几位著名设计师的共同创作亦是酒店吸引人的重要原因。

墙上的花园

被钢筋混凝土推着跑的城市，偶尔需要停下来，呼吸一下大自然绿草如茵的清凉味道。而如若这芳草萋萋，变成一幅画怎样？建筑师Patrick Blanc就为hotel icon带来了这样的震撼。他将一整片绿植围合成的不规则花园，垂直悬挂在酒店前台背后的墙体上。这片巨大的立体花园将酒店一二层空间垂直串联起来，形成一处大型的天然背景，与大堂入口处的木质半围合状态的座椅相互呼应，静坐期间，无不感觉身心放松。

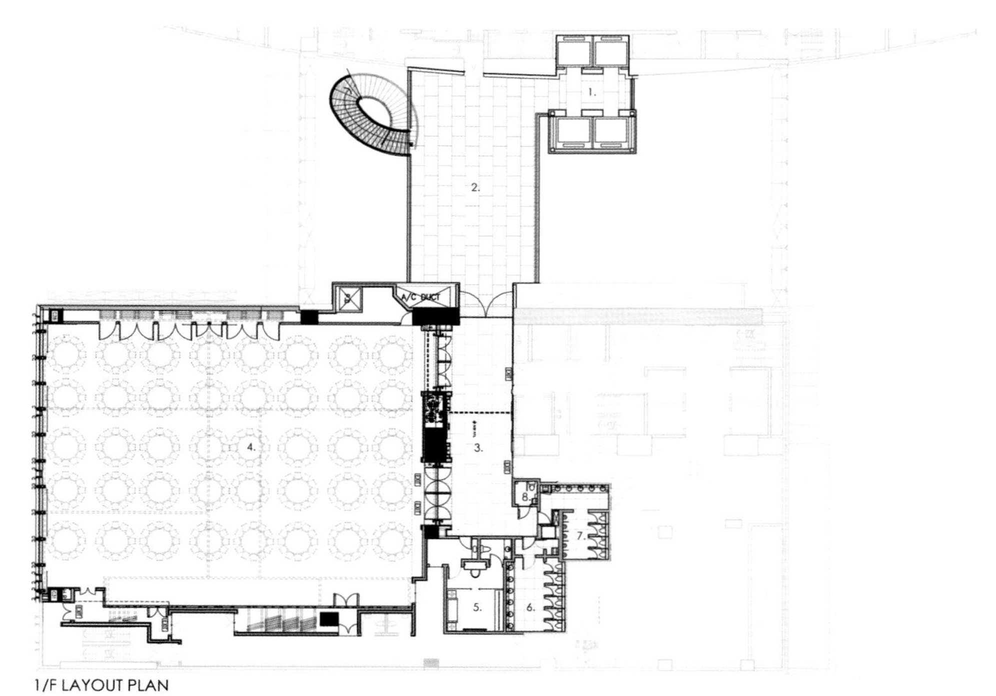

1/F LAYOUT PLAN
1. LIFT LOBBY
2. LINK BRIDGE
3. PRE-FUNCTION AREA
4. BALLROOM
5. BRIDAL ROOM
6. FEMALE LAVATORY
7. MALE LAVATORY
8. DISABLED LAVATORY

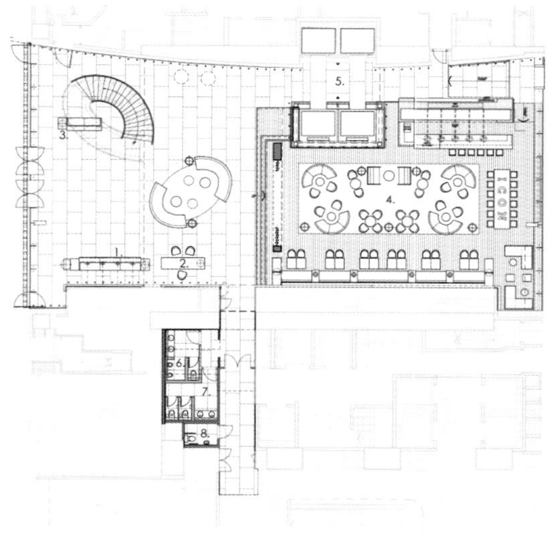

G/F LAYOUT PLAN

1. RECEPTION
2. GUEST RELATION
3. CONCIERGE
4. LOUNGE
5. LIFT LOBBY
6. MALE LAVATORY
7. FEMALE LAVATORY
8. DISABLED LAVATORY

平衡之美

酒店不论从功能布局还是材质运用上都蕴藏了中国的道家阴阳之说。学说认为自然界任何事物或现象都包含着既相互对立，又互根互用的阴阳两个方面，之间消长平衡的关系是本次设计所想体现的主题，有方即有园，有刚亦有柔，大堂内家具柔和的曲线、以及以单纯的层压板和内部照明制作出连接二层的回旋向上的楼梯均中和了建筑的方正棱角，极具雕塑形态。同时这些木质家具、楼梯也从材质和色彩上平衡了酒店公共空间中被大量使用到的大理石、玻璃、钢筋等。就连客房中也大量使用了曲线元素，如弧形的浴室门。

老香港的回忆

从入口径直往里走，拾阶而上，是酒店一处带休闲餐饮功能的休闲大堂。而连接两个空间的是一幅巨型门套。其黑色镂空的连续图案、可以伸缩折叠的铁闸门，瞬间将许多本地顾客的思绪拉回到儿时的香港，那铁门因生锈而产生的犀利声响，连同回忆都渐渐清晰起来……这也是很多本地顾客喜欢这里的原因之一，不能不说是设计师的心思细腻。

铁门旁的玻璃观光梯塔，则通过文字的拆解、沉积，用现代的设计手法形象的展现了中国古老的文字艺术。

“明亮”的诱惑

酒店照明是一个庞大而复杂的系统，不仅要考虑灯具的美观、耐用、省电节能，还要把控包括眩光影响、光源收藏、氛围营造等一系列问题。Hotel icon的照明尤其是节能以及环境营造方面绝对是一个值得参详的例子。

酒店宴会厅，是一处以主要承办宴请及会议的场所，为了跳脱常规照明方式的桎梏，凸现惊艳之感，设计将定制的菱柱形水晶玻璃天花代替了传统的吊灯。巨大的矩形天花，如璀璨的天幕，淡紫色的灯光，明亮而不刺眼，营造出高贵浪漫的氛围。

另外，一楼前台的柱形吊灯，也很好地解决了由于大堂层高而造成的照明的问题。当然最有意思的还是酒店大厅东面吧台上摆放的白色台灯。台灯由设计师定制，是酒店名称的造型，当灯光投影到桌面，便自然形成单词icon的影像，光源柔和，饶有趣味。

透过健身中心巨大的落地玻璃，窗外是夜晚维港的灯火阑珊。美景如斯，让人流连忘返！

The Hong Kong Hotel Icon locates at the main tour district of Kowloon, is a teaching and research hotel established by Hong Kong Polytechnic University, has 262 rooms, and is a hotel which can view the Victoria Harbor seascape. Beside the perfect geographical location, the cooperation creation of several designers is also a main reason in attracting the guests.

Garden on Walls

The city which is pushed to run by the steel-reinforced concrete should occasionally stop to breathe cool and refreshing flavor of natural grass. If the grass is luxuriant, how about it becomes a picture? The architect Patrick Blanc brings such shake for Hotel Icon. He encloses the whole green plants into an irregular garden to vertically hang onto wall behind hotel front-desk. Such large 3D garden vertically connects the spaces of hotel 1st and 2nd floors, to form a large-scale natural background; it corresponds with the semi-closed seats at lobby entrance; seat in such environment, your will feel relaxation of mind and body.

Beauty of Balance:

The functional arrangement and material utilization of such hotel contain the Yin & Yang concept of Taoism. The Taoism thinks that every thing or phenomenon in natural world contains two aspects which are opposite, and root & utilize from each other; the growth & decline balance relationship between two aspects are the main themes of this design; if there is the square, there must be the circle; if there is the rigid, there must be the flexible; the gentle curves of lobby furniture, the pure laminated boards and internal lighting produce the circling-upward stairs to connect the 2nd floor, these balance out the square edges and corners of building, and have the sculpture shapes. And, the wooden furniture and stairs balance out the marbles, glass, steel bars and others utilized in hotel public space, from material and color. Even the guest rooms also utilize abundant curve elements, such as the arched bathroom door.

Memory of Old Hong Kong

Directly walk into the hotel from entrance, and walk upwards step by step, there is a leisure lobby with leisure and dinning functions. A large-scale door pocket connects the two spaces. The black hollowed-out continuous figures and the flexible and foldable steel gate can instantly draw the may local guests' thinking to childhood Hong King; the sharp sound created by rusted steel gate and also the memory become more and more clear …… this is a reason that many local guests love here. It is the success of fine ideas of designers.

The glass sightseeing ladder tower next to steel gate shows the China ancient character art with character disassembly, deposition and modern design methods.

Attraction of "Brightness"

The hotel lighting is a large and complicated system, should consider about the lamp beauty, durability and power-saving, and also control the glare influence, light-source hiding, atmosphere creation and other issues. The lighting of Hotel Icon, especially the power-saving and atmosphere creation, is a valuable reference example.

The hotel banquet hall is a location to hold entertainments and meetings, in order to avoid the limitation of normal lighting and protrude the gorgeous feeling, the design use the customized prism-shape crystal-glass ceiling instead of traditional ceiling lamp. The giant rectangle ceiling likes the bright heaven canopy; the lilac light is bright but not dazzling, and creates noble and romantic atmosphere.

And, the column-type ceiling lamp at ground-floor front desk also solves the lighting issue caused by high lobby. The most interesting is the white desk lamp on the bar counter at east of hotel lobby. Such lamp is customized by the designer, has the shape of hotel name; when the lamp light projects onto the counter surface, the word "ICON" is naturally formed; the light source is gentle and interesting.

Through the large full-height glass of health-building center, the outside is the light waning of night Victoria Harbor., the people enjoy themselves in such a great scene and forget to go back.

奕居
THE UPPER HOUSE

设计　Andre Fu

设计单位：AFSO
项目地点：香港

太古酒店旗下提供个性化亲切服务的精致豪华酒店——奕居座落于香港太古广场之上，以当代永恒的亚洲设计风格，为宾客呈献恬静的空间。

由Andre设计的奕居酒店巧妙地利用天然素材、独特的雕刻品及原创艺术品，打造当代亚洲设计风格，以配合奕居宁静居所的气氛。他表示："我刻意打造和谐恬静的酒店设计及环境，营造犹如私人居所般的个性化精致豪华酒店。"

奕居总经理Dean Winter补充说："奕居邀得Andre Fu主理酒店设计，为宾客于香港闹市之中呈献远离烦嚣的住宿享受。"

Thomas Heatherwick设计的Bedonia石制外墙入口，一如敞开门帘般欢迎宾客亲临此私人居所——奕居。宾客穿过设有灯光效果的车道后，甫见到夺目的4 m高镀镍大门。大门四周被落地玻璃所环绕，犹如悬浮于水上一样。

宾客踏进酒店入口之后，通过环型竹子构成的“The Lantern”展开登上恬静舒适的旅程。电梯渗透柔和灯光，而墙上则悬挂着独一无二的灯饰，引领宾客登往6楼“The Lawn”。此乃茂密的室外草坪，气氛轻松，让宾客沐浴于自然阳光之际，享用鸡尾酒。

酒店设有117间宽敞客房，包括21间套房及2间顶层套房，所有客房均以天然木、日式透光玻璃、石灰岩地砖及墙砖、漆面墙纸等原料打造摩登的单色设计，不经意流露着低调奢华风格。此外，所有客房更以两大颜色为设计主题：“竹子Bamboo”，房间铺设实灰色地板、并饰以竹子及淡紫色地毯；“青瓷Celadon”，房间铺设绿茶色地毯、橡木地板，并饰以米色橡木做装饰。

客房空间宽阔，而浴室面积更达30 m^2，可饱览窗外醉人海景或优雅城市景观。浴室铺设意大利Perlato Svevo天然石灰岩地砖，墙身则为土耳其Terre d'Oriente米色石灰岩，设计简约优雅，并配备步入式淋浴间、梳妆区及深浸浴缸。

酒店的焦点为49楼的Sky Bridge，

展现卓越的建筑设计。Sky Bridge 位于 40 m 高的中庭之上，并设有天窗，引领宾客进入 Sky Lounge 及酒店的现代雅致餐厅及酒吧 Café Gray Deluxe。

酒廊中央设有壁炉，气氛温馨亲切，全日提供多款餐饮及鸡尾酒。酒廊设有 4 m 特高天花，让宾客躺卧于绿茶色及淡蓝色的沙发中享受美酒。

酒店的灵感源自旧巴黎豪华咖啡厅的现代雅致餐厅 Café Gray Deluxe，坐拥维港迷人景色，由重临香江的国际名厨 Gray Kunz 主理。Café Gray Deluxe 设计充满活力，设有 14 m 长开放式厨房及吧台。Gray 选用了本地有机食材炮制各款滋味美食，配合餐厅悠闲的服务态度，为宾客呈献非一般的日常佳肴。

主用膳厅环抱维港迷人美景，可容纳约 100 名宾客；而半开放式用膳区的宾客则可欣赏开放式厨房内厨师的烹调过程。此外，Café Gray Deluxe 的私人宴会厅尽览醉人海景，可招待 12 人；时尚设计及充满生气的酒廊更可容纳多达 88 人。

酒店内展示多件卓越的当代艺术品，包括利用砂岩、陶瓷、云石及青铜制成的雕塑品，完美地配合 Andre Fu 营造的恬静设计及气氛。

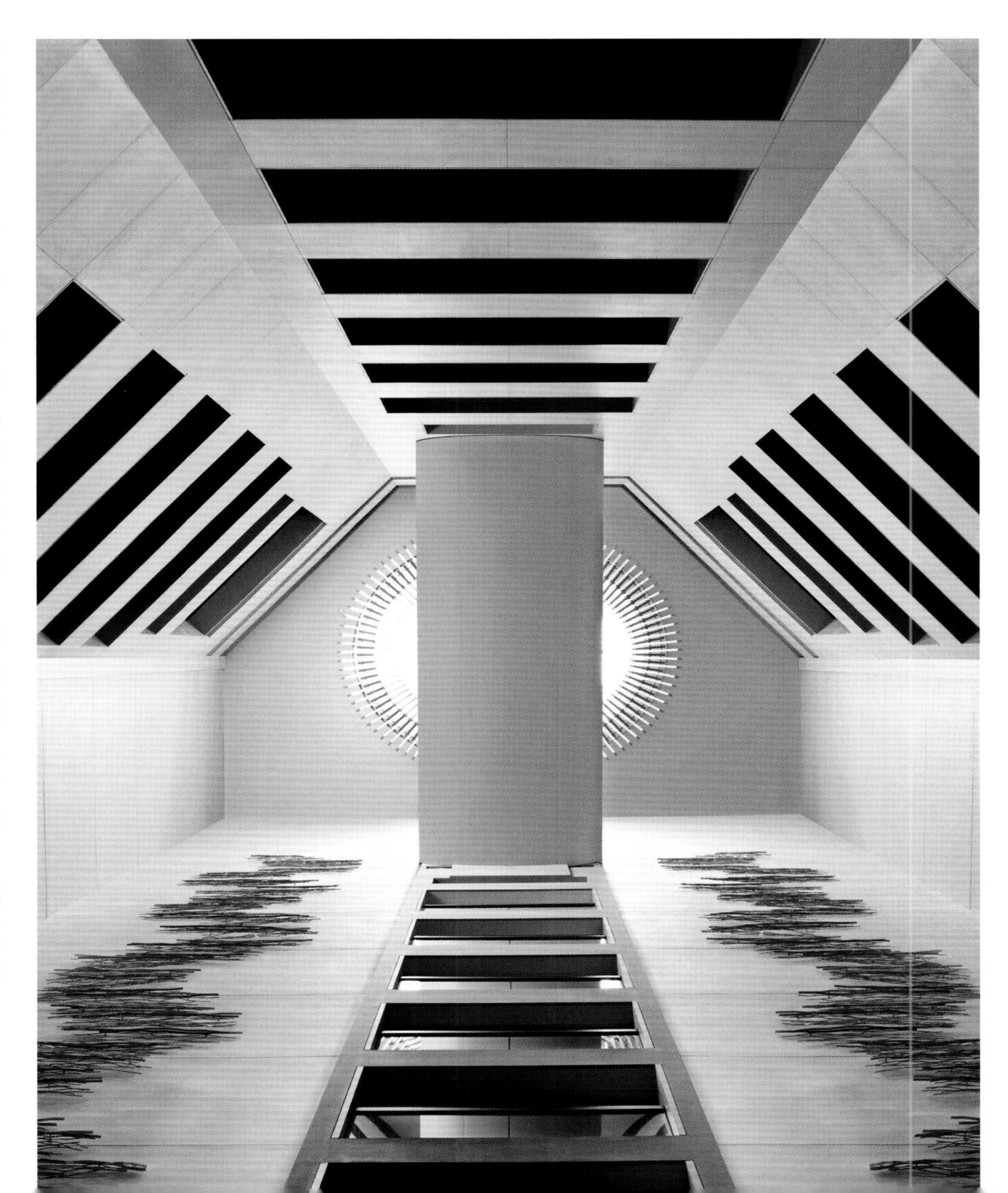

Affiliated with Taigu Hotel, there is fine and luxury hotel that provides co-nial services—Yiju. Yiju is located at Taigu Square of Hong Kong. It is adopting the classic Asia design style and providing a peaceful space for people.

Designed by Andre, Yiju Hotel has adopted natural materials, unique carvings and original arts cleverly, which have worked out the modern Asia design style so as to suit for the quiet space. This hotel, aiming to create a harmonious and peace in terms of design and environment of hotel, provides people a private residence but exquisite and luxury. It will give guest a comfortable and peaceful feeling even in the hustle and bustle Hong Kong."

The Bedonia stone entrance of the wall, designed by Thomas Heatherwick, like a private residence to welcome guests—that' s Yiju. Passing through the lighting lane, there is an eye-catching nickel-plate door. This door is surrounded with glass, liking suspended on water.

Entering the hotel, guests will have a great trip through "The Lan-

tern" consisted of round bamboos. There is soft and warm light spreading in the elevator. Furthermore, "The Lawn", leading guests to the 6th floor, is hanging on the wall, which is decorated with the only and unique lighting. This is a wonderful lawn. Guests could bath in sunshine and enjoy cocktail for relax.

This hotel has 117 spacious rooms, including 21 suites and two top class suits. All the rooms are furnished in natural wood, Japanese-style translucent glass, limestone floor tiles and wall tiles, paint wallpaper and other materials to create a modern monochrome design, low-key but luxury. In addition, all rooms are adopting two major colors as the design theme: one is "Bamboo". The house is laying a gray color floor but decorated with bamboo carpet. The other is Celadon. The house is laying green-tea color carpet, oak floor but decorated with beige oak.

The rooms are spacious, and the bathroom area is nearly 30 square meters even, which can provide you charming sea views or elegant urban landscape. The bathroom is adopted Perlato Svevo, a natural limestone floor tiles of Italy. The walls are adopted Terre d' Oriente beige limestone of Turkish. The design is simple but elegant, which is equipped with walk-in shower, dressing area and deep soak-

ing tub.

The most shining point of this hotel is the Sky Bridge in the 49th floor, showing its excellent architectural design. Sky Bridge is located above the 40-meter-high atrium and is set sky window, leading guest enter the Sky Lounge and the hotel's modern and elegant restaurant and bar named Café Gray Deluxe.

There is a fireplace in the central lounge, warm and charming, providing a variety of dining and cocktails all day. The ceiling of this lounge is four meters high, allowing guests enjoying the wine by lying in the green and blue sofas.

The inspirit of this hotel comes from the modern and elegant Café Gray Deluxe of Paris, bordering the charming landscape. The chief is the assistant of international famous Gray Kunz. The design of Café Gray Deluxe is active very much. There is a fourteen-meter-long open kitchen and bar. Gray selects the organic food to cook different tasty food. Coupled

with the nice service attitude, guest will enjoy a wonderful daily food.

The major dinning room is surrounding the charming landscape, which can host about 100 guests. Furthermore, the semi-open dining area could enjoy the cooking process in the kitchen. In addition, the private dining room of Café Gray Deluxe offer guest a beautiful and charming view, which could host 12 guests. The modern design and the lively lunge can even host 88 guests.

There are demonstrating many great modern arts, including the sculptures made of sandstone, ceramic, marble and bronze, which has echoed with the peaceful design atmosphere created by Andre Fu perfectly.

<4701-4707

上海外滩英迪格酒店

INDIGO'S SHANGHAI ON THE BUND

设计 HBA

设 计 师:Andrew Moore
项目地点:上海
主要材料:原钢、混凝土、外露砖、
抛光石膏灰砖、灰色嵌板、
抛光石膏墙、帆布

由享誉国际的室内设计公司Hirsch Bedner Associates(HBA)一手打造的上海外滩英迪格酒店(Hotel Indigo Shanghai on the Bund)最近惊艳亮相上海。该酒店是洲际酒店集团旗下亚洲首家英迪格酒店,HBA的创新设计兼收并蓄而又亲切和谐,体现了上海东西交融、海纳百川、面向未来的城市精神。

英迪格品牌的理念是呈献融汇本土特色的精品酒店,让宾客产生与当地社区紧密相连的亲切感。HBA为了使这家英迪格亚洲旗舰酒店实现这一愿景,精心打造出了一家"拥有独特个性"的酒店,设计遍及酒店的180间客房,当中包括21间江景套房及两间宽敞的花园露台套房。

酒店的大堂入口异常绚烂瑰丽,堪称沪上一绝;既反映了酒店位于黄浦江畔的位置,还体现了品牌对自然环境、循环再用,以及生态敏感型设计的承诺。HBA选择原钢、混凝土、外露砖及抛光石膏等富有张力的基本材料为大堂进行装璜,令人不禁联想到这一空间是从码头旁的滨江阁楼改建而来。而开放式隔室与清水混凝土天花便进一步增强这种效果,并配以全日色彩幻变的灯光。

与大堂如出一辙，客房也呈现一种自然色调：外露的上海灰砖、磨耗效果的灰色嵌板、抛光石膏墙和帆布。与之产生强烈对比效果的是色彩鲜艳跳跃的地毯。

中式灯笼、传统家具、陶瓷和古董等兼收并蓄、机巧别致的工艺品和家具带来老上海的感觉。带顶篷的睡床为原创设计，灵感源自传统中式婚礼所用的喜床，经当代手法重新演绎。

偌大的浴室设有一堵镶在抛光钢框中的玻璃墙，望向黄浦江；并设开放式湿区，当中附设配上长方形瓷面盆的简约盥洗台，营造当代风尚；而独立浴缸也同样时尚摩登。

HBA 这一创新设计古今交织，堪称奇迹。呈现在世人面前的是一个充满年轻活力、符合当代潮流、蕴含无限灵感的极致空间。这一空间从上海的历史走来，并将开创上海未来设计的新风尚。

HBA 设计的这一亚洲首家英迪格酒店，使英迪格品牌雄踞上海外滩十六铺这一充满近现代历史风云的时尚新地标，并且为未来的英迪格酒店树立了一个可以借鉴的标杆。

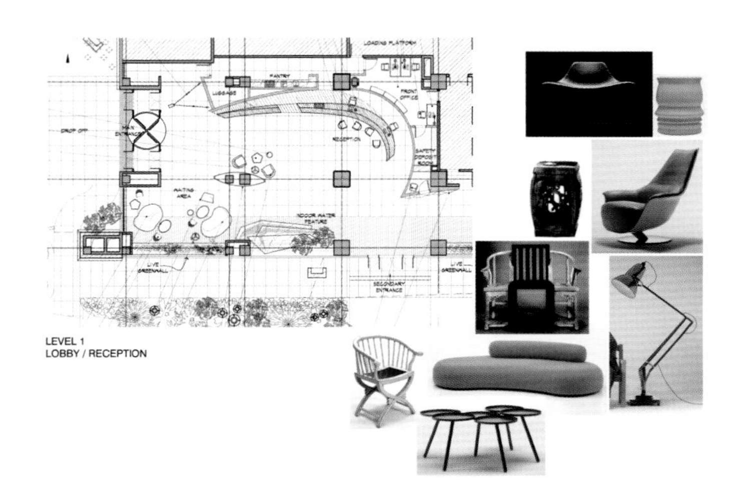
LEVEL 1
LOBBY / RECEPTION

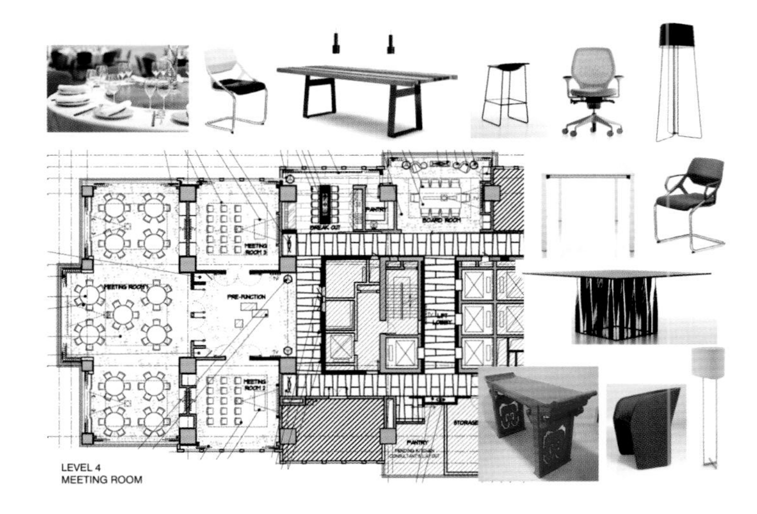
LEVEL 4
MEETING ROOM

With the goal of transforming Hotel Indigo Shanghai on the Bund, flagship for the debut of InterContinental Hotel Group's boutique Hotel Indigo brand in Asia, global interior design firm Hirsch Bedner Associates (HBA) has stamped an innovative design onto the hotel that is at once eclectic and harmonious, a design that connects the ancient with the modern.

The philosophy behind the Hotel Indigo brand is to offer boutique-style hotels infused with local inspiration so guests feel connected to the local neighbourhood and community. To fulfil this vision at the Asian flagship, HBA's concept was to create a "personality all its own" for the 180-room hotel, including 21 River View Suites and two spacious Garden Terrace Suites.

"Throughout, Hotel Indigo's design is about connecting the hotel to the neighbourhood—one anchored by the river and its influence on commerce and connection," said Andrew Moore, HBA's lead designer on the project.

HBA developed an eclectic and harmonious design linked to the neighbouring Huangpu River, and the element that ties it to the neighborhood most intimately, Shiliupu Dock, now know as Pier 16. This dock was the gateway through which Shanghai grew, as a shipping and trade centre, and entry point for thousands of European expatriates who led Shanghai's development as a global city.

"While a general sense of place is often a hallmark of thoughtful hospitality design, HBA dials in place at a whole new level of detail for the Hotel Indigo: not a nation or even a city, but a neighbourhood," added Mr Moore.

The lobby entrance is among the most striking and dramatic in Shanghai, reflecting Hotel Indigo's position on the river and the brand's commitment to nature, recyclables and ecologically sensitive design.

HBA chose strong elemental materials to render the lobby: raw steel, concrete, exposed brick, and polished plaster – suggesting this gallery space has been repurposed from a wharf-side waterfront loft. The open cell, cast concrete ceiling enhances this effect, studded with lighting that changes colours throughout the day.

As in the lobby, the guestroom palette is the natural tone of exposed Shanghai gray brick, distressed gray paneling, and polished plaster walls, a canvas against which shines colorful and lively carpets.

The sense of an older Shanghai is in eclectic and whimsical artifacts and furniture: with Chinese lanterns, authentic furniture, ceramic pieces and

antiques. The canopy bed, an original design, was inspired by traditional Chinese wedding beds, but reinterpreted though a contemporary lens.

"We found wonderful furniture—for example, an especially interesting armoire console—in the local bazaars," said Mr Moore. "We had it restored, then sprayed in fresh white enamel, so the piece would be simultaneously old and new." It became the model for reproductions used in each room. Other furnishings reflect ecological sensitivity: while the pieces vary room to room, the materials are all eco-friendly.

Oversized bathrooms have a glass wall framed in polished steel, looking out onto the river. They feature an open wet area, where a minimalist vanity topped with rectangular porcelain basins gives a contemporary feel, as does the freestanding tub, which is a sleek and modern.

HBA's innovative design accomplished a true feat: connecting the ancient with the modern. The result is a youthful, contemporary, inspired space that understands where it has come from and leads the way into Shanghai's design future.

In helping develop the first Hotel Indigo in Asia, HBA created a design that transforms the boutique brand name into a vantage point onto the most interesting areas within the storied and dynamic city of Shanghai – creating a standard against which all future Hotel Indigo properties will be judged.

AURORA

A LAST LOOK

BUILDING SHANGHAI

citizenM 酒店

CITIZENM HOTEL BANKSIDE LONDON, UK

设计　concrete

参与设计：Rob Wagemans、Erikjan Vermeulen、Michael Woodford、Jurjen van Hulzen、Sander Vredeveld、MarcBrummelhuis、Menno Baas
建筑面积：5800 m²
完工时间：2012
摄 影 师：Richard Powers

citizenM是一家新荷兰集团，他们于2008年在史基浦机场开业了第一家酒店。citizenM Bankside将是第4家开业酒店，为全球移动居民提供位于城市心脏地带可支付得起的奢华享受。这家酒店的概念要求去除所有隐性成本以及所有非必要项目，从而为客人提供预算价格内的奢华感受。该酒店具有192间客房，每间14 m²；所有预制件产自一家工厂，运输方便。设计是基于citizenM信念：我们在城市或商务旅行期间所需要的就是一张好床以及简单清洁的卫生间。这些房间堆叠在第一层，具有动态的大厅、活动客房空间以及F&B功能区，包括一个公共咖啡厅。7个"创造性"空间位于第一层，是以societyM之名运营，也就是citizenM的工作和会议设施。

建筑位于Southwark邻近地区。该区域传统上主要是工业仓库。本建筑物使用工业仓库作为参考，通过使用之前用于修筑仓库的沙色伦敦规格砖块重新演绎的颜色与材料来参照过去。模块化修建的citizenM酒店通过仿混凝土板的定制砖块仔细覆盖。

在该酒店内，与艺术的关联比之前酒店要重要得多。不仅仅因为该酒店位于Tate Modern附近，更因为艺术给予客人另一种看世界的可能性。不仅在酒店内部，还包括在酒店外部，艺术给予

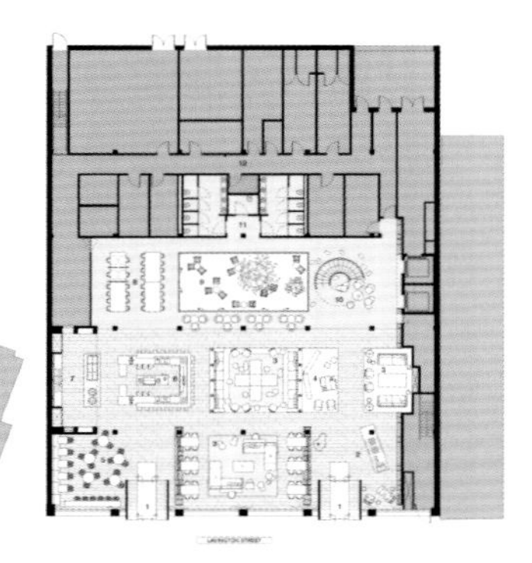

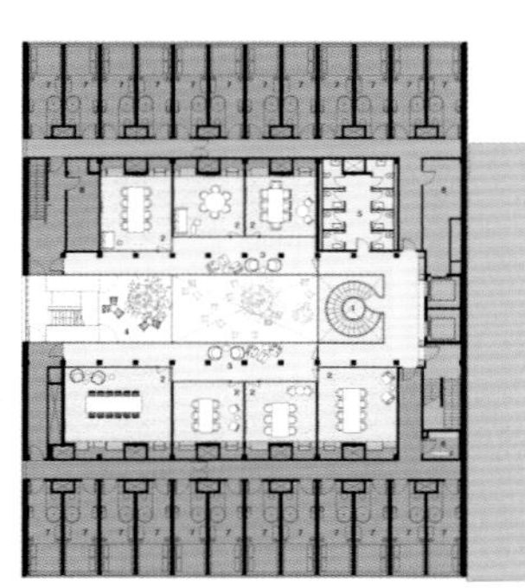

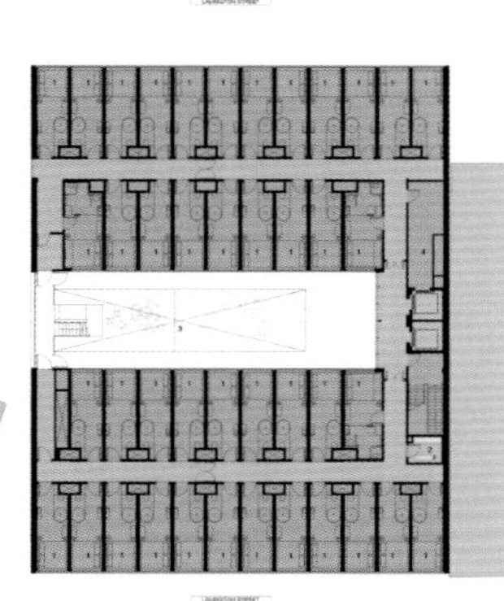

citizenM顾客以及伦敦人作出反应的机会，以及给予艺术家展示的平台。citizenM岸边酒店展示了Mark Titchner令人惊讶的新作品，称为“另一个世界是可能的”。

该地点提供了安置相互交错两栋客房的可能性，使得第一栋citizenM有庭院。庭院设计成户外客厅，是这家新酒店的精髓部分。庭院在酒店的中心地带形成美丽的绿洲，将日光引入客房、societyM以及酒店的公共生活区。多个楼层上具有阳台，您可以在上面喝酒，享受新鲜空气，以及吸烟——如果您还有这个爱好的话。具有特殊设计砌砖图案的沙色石板与相邻的仓库建筑相呼应。在这个巨大的空间，客房的大玻璃窗被向外推出。这些铝框的各种深度具有独特扭曲的形态。地面上大型的玻璃窗置于建筑砌块之内，形成了内外的自然转换，清晰展示了朝向大街的客厅、咖啡厅和大厅充满生气的生活。木质进门门套清楚指出了朝向咖啡厅以及酒店大厅的入口。

citizenM is a new Dutch hotel group that opened their first hotel at Schiphol Airport in 2008. citizenM Bankside will be the fourth hotel to open, and offers mobile citizens of the world affordable luxury in the heart of the city. The concept of the hotel is to cut out all hidden costs and remove all unnecessary items, in order to provide its guests a luxury feel for a budget price. The hotel exists of 192 rooms of 14 sq m, all prefabricated produced in a factory and easy to transport. The design is focussed on citizenM's belief that a great bed and a simple and clean bathroom is all we need during a city or business trip. The rooms are stacked on a ground floor with a dynamic lobby, living room space and F&B functions including a public accessible cafe. Seven ' creating' spaces are housed on the first floor, operating under the name of societyM, citizenM's working and meeting facility.

the building: The building has landed in the upcoming neighborhood of Southwark. An area that traditionally has been dominated by indus-

trial warehouses. This reference was used to create a building that refers to the past by using a colour and material that is a reinterpretation of the sand coloured London stock bricks, formerly used to construct warehouses. This modular build citizenM hotel is carefully covered with custom- made brick patterned concrete panels.

Within this hotel the connection with art is much more important than before. Not only because it's located close to the Tate Modern, but also because art gives guests the possibility to look in another way at the world. Not just on the inside of the hotel but also on the outside of the building. This gives not only citizenM's guests, but also Londoners the possibility to have this moment of reflection and give the artist a great platform to perform. The

citizen
welcome citizen
5118-TAT
NEW YORK

citizenM Bankside hotel displays an amazing new work by Mark Titchner called Another world is possible.

The location provided the possibility to position two blocks of rooms behind each other, which results in the first citizenM with a courtyard. This courtyard is designed as outdoor living room, so it becomes an essential part of the experience of this new hotel. The courtyard creates a beautiful oasis in the heart of the hotel and brings?daylight into the rooms, societyM and the public life of the hotel. There are terraces on several floors that can be used for drinks, some fresh air, or a smoke if you still have this habit. the facade: Sand coloured stone panels with special designed brickwork pattern are referring to the warehouse architecture of the neighbourhood. Within this robust volume big glass windows of the rooms are pushed out. The various depths of these aluminium frames give an individual twist to the rigid. The large glass windows on the ground are placed within the building block, creating a natural transfer between inside and outside. They clearly show the vibrant life of the living rooms, cafe and lobby facing space towards the street. Two wooden entrance boxes clearly indicate the entrances towards the café and the hotel lobby.

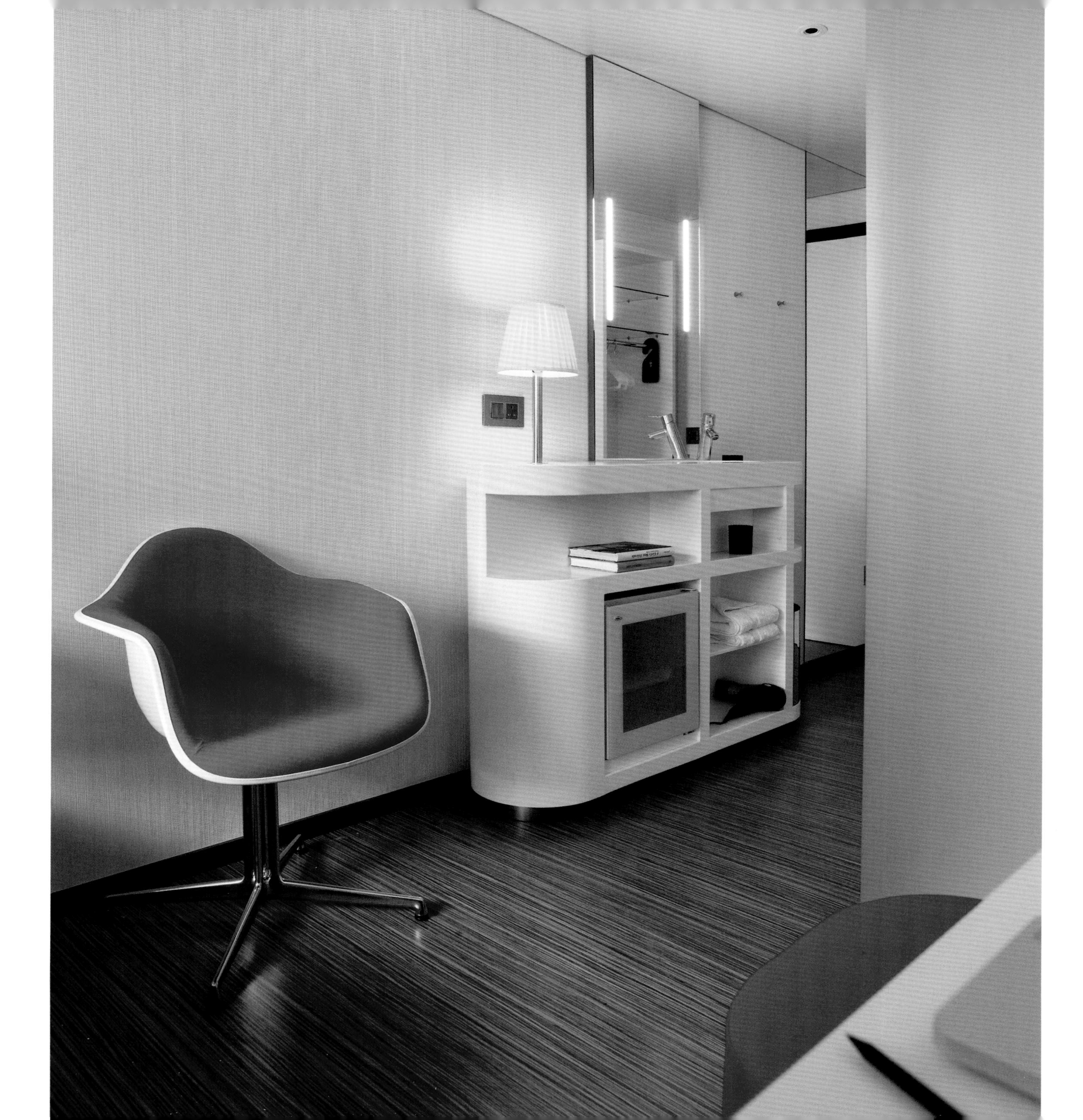

成都岷山饭店

CHENGDU MINSHAN HOTEL

设计　杨邦胜

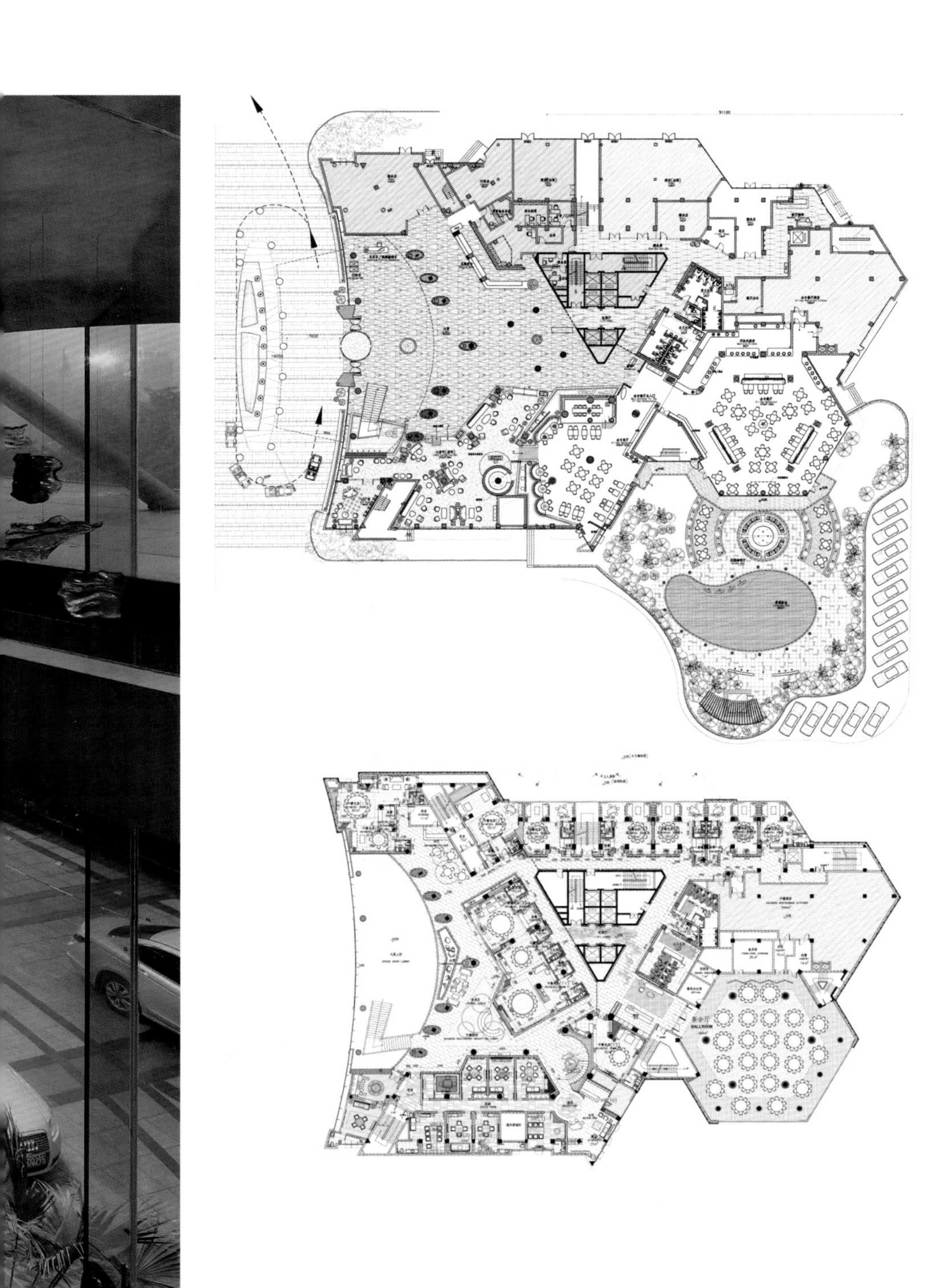

该作品为翻新改造设计项目。建于上世纪80年代的成都岷山饭店，是西南地区最早的高星级酒店，由于年岁久远，亟需翻新改造。设计中，岷山饭店被重新定位为精品设计型城市商务酒店，设计紧扣川蜀地域文化特色，从川蜀文化中提取休闲文化、岷山文化等设计元素，并将其化作酒店内上百只倒扣茶碗吊顶、黑白分明的毛笔灯饰群、随处可见的岷山水墨图案，成都的"市花"芙蓉也变成了大堂空中悬落的芙蓉花瓣灯饰，呈现一幅隽永的川蜀水墨画卷。作品运用国际化的设计理念和表达方式，使改造后的岷山饭店重现昔日的辉煌与荣耀。

设计单位：YAC（国际）杨邦胜酒店设计顾问公司
项目地点：四川省成都市人民南路二段55号
建筑面积：2.5万 m^2
主要材料：米黄石、金属、琉璃、玻璃、仿古砖
摄 影 师：贾方、马晓春

This piece of work is a re-shaped design project. Built in the 1980s, Chengdu Minshan Hotel is the earliest star hotel in southwest. Due to long history, then it need to be re-shaped once immediately. In the design, Minshan Hotel is re-positioned as a boutique urban business hotel. This design is close to the local culture, such as leisure culture and Minshan culure extracted form Sichuan province culture. In addition, the designer has converted them into hundreds of inverted bowls on the ceiling. They black and white color lights, Minshan Mountain ink pictures can be seen easily. The municipal flower of Chengdu city Furong Flower has become a Flos Hibisci Mutabilis Light hanging in the lobby, showing a meaningful Sichuan ink volume. This piece of work adopts the international design concept and expression way and makes this re-shaped Minshan Hotel show its former glory and gorgeous.

东方商旅

LES SUITES ORIENT, BUND SHANGHAI

设计　吴宗岳

设计单位：吴宗岳设计有限公司
项目地点：上海市黄浦区金陵东路一号
建筑面积：16986 m^2
主要材料：大理石、深色实木

东方商旅酒店位于南北外滩交界的中心点，对望浦东陆家嘴，后连豫园及十六铺码头观光区。楼高23层，外观为20世纪30年代装饰性艺术(Art Deco)的风格，邻近“万国建筑博物馆”，且曾亲历并见证了中国航运发展与港口建设的历史。该建筑前身曾是19世纪60年代由美国人罗赛尔设立的旗昌洋行，罗赛尔并建设了利源码头，也就是现在十六铺码头的前身。

延续台北商旅(Les Suites Taipei)低调奢华的风格，酒店的设计将过去上海滩的灵感与上个世纪30年代装饰性艺术元素优雅融合。

拥有168间豪华客房，包括43间套房，是唯一能以270°视野尽览北外滩、南外滩、浦东、浦西江景的景观酒店，深刻感受浦西史迹与浦东摩登时空交错，相互辉映。

酒店入口低调，充分注重隐密性。踏上灰色石瓦地，推转金铜旋转门，一楼接待处的大理石散发出温暖的鹅白色。在黑色大理石、深色实木、古铜色铜雕、白色陶胚构成的质感讲究的空间里，感受静谧与温馨的服务。

客房内装精心设计，自然质朴的成色，柔和宁谧的灯光，为身处喧嚣城市中的住客营造居家的温馨感觉。所有客房里的家具和文具都是特别订制，大地色系中多使用了一点灰，让家具不只是舒适，更有一种诚恳而内敛的气质。收纳式的迷你多宝阁整齐地收纳着咖啡机、各式调酒、茶壶组与玻璃杯。浴室镜子中，有一小面显示时间和当日气候与气温的屏幕，贴心地让房客知晓室外天气状况。此外，酒店的房卡兼备有上海交通卡功能，可以乘坐计程车、地铁、甚至渡轮，是上海第一家使用此智能、贴心功能的饭店。酒店更为住客提供商旅智能生活手机，作为客人在当地使用的手机，方便所有商务旅客，将酒店的服务无限延伸。

酒店内规划有各式多功能活动区，根据客人的需求，量身打造精致化的服务内容，助您完美呈现每一场活动。除为繁忙的商务住客提供私密的独立会议室，具延展性且设备齐全的空间亦可作为记者会、产品展示会或是时尚发表会使用。

Located in the center of the place where South and North Bunds join, Les Suites Orient is opposite to Lujiazui Pudong, with its back leading to Yuyuan and Shiliupu Wharf, a sightseeing place. The hotel is a 23-storey building, with its appearance styled in Art Deco, which was common in 1930s. Adjacent to the buildings from different countries, it has experienced and witnesses the history of China's shipping industry and harbor construction. The building of the hotel used to be the building of Russel & Co., which was established in 1860s by Russel, an American, who also start Liyuan Wharf, the ancestor of Shiliupu Wharf

Inheriting the style of Les Suites Taipei, which appears low-profile but actually luxurious, the hotel was designed by merging the inspirations from Shanghai Bunds with the elements of Art Deco common in 1930s.

With 168 guest rooms, including 43 suites, the hotel is the only sightseeing hotel which allows you to view all the sights in North and South Bunds, east and west banks of Huangpujiang River in 270° field of vision, to feel the relics in west bank and modernism in east bank of Huangpujiang River.

The entrance to the hotel is low-profile, with full attention to privacy. The grayly tiled floor and the bronzed revolving door lead to the information desk on the first floor with the marble as white as the color of geese, which look so warm. The guests can feel the quite and gentle service in a space with black marbles, dark real wood, bronzed statues and white potteries, which provides sense of reality.

The interiors of the guest rooms have been carefully designed, with natural and plain color, gentle and inconspicuous lights, to provide the guests in a busy city a sweet home feelings. All the furniture and stationery

in the guest rooms are custom, with earth-colored paint accompanied by a little gray, so that the furniture looks not only cozy,but also sincere and low-profile. In the mini cupboard for storing are the coffee maker, different wineglasses, cups and a kettle. In the mirror in the bathroom, there is small display showing the local time,weather and temperature, which considerately allow the guest to know the weather outdoor. In addition, the room cards of the hotel can be used as transportation cards in Shanghai, which can be used to take taxi, subway and even ferry ship, which is the first in Shanghai. Besides, the hotel provides everyone of its guests with an intelligent mobile phone usable locally for business life and travel, to makes easy all the guests and extend the service of the hotel.

Different multifunctional areas are planned in the hotel to respond to different needs of its customers and the hotel can provide customized service to help you perfectly present each activity. The independent meeting rooms can not only be served for the privacy of business guests, but also for press conference , product exhibitions and fashion shows with extendability and complete equipment.

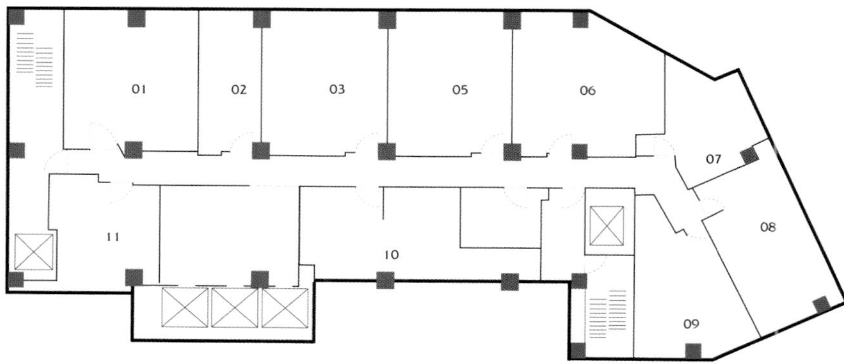
01
02
03
05
06
07
08
11
10
09

苏梅岛 W 度假酒店

W HOTEL KOH SAMUI

设计　MAPS Design

MAPS Design设计的苏梅岛W度假酒店，位于泰国苏梅岛湄南海滩，是东南亚第一家W度假酒店。这家别墅式海滨度假胜地占地约17300 m²，共有70栋别墅，每栋别墅约160 m²，拥有W度假酒店的特色客厅和酒吧间以及Spa。这里还有17处私人住宅，每栋面积为700~1300 m²，占地面积在1000~2500 m²，住宅内部可远眺全部海景。私人住宅区建在一处约900 ha的隐蔽海滨。

项目地点：泰国
建筑面积：17300 m²

与 MAPS 设计工作室通力协作，并配以由著名泰国室内设计公司 P49 Deesign 精心打造的迷人亮点，让客人全天沉醉在精致优雅的度假生活氛围中，感受无与伦比的休憩、放松、恢复和新生。诚邀您前来大快朵颐厨房餐桌的全球美食，或在酒店最令人惊叹的布景——“仙境边缘”品尝 Namu 餐厅创意十足的日式料理。来到 W 海滩边放松小憩、静心阅读，让我们的海滩大使为您奉上各种贴心服务，还可在 WAVE 海浪中尽情体验花样繁多的水上活动。前往我们的逍遥之地 AWAY 水疗中心，利用各式舒缓身心的天然疗法恢复您的感官平衡，同时尝试一下亚洲酒店业内唯一的墨西哥 Temascal 特色疗法——Thaimazcal。还可在设施一流的 SWEAT 健身中心强身健体，体验焕发活力的网球、泰拳、瑜伽和健美操运动。或在我们的 WET 泳池中尽情畅游一番，感受清凉舒爽的奢华体验。

W

Designed by MAPS, Koh Samui W Resort Hotel, located at the Maenam Beach, Thailand, is the first-ever W resort hotel in Southeast Asia. This villa-style seaside resort was built at the Maenam Beach, north coast of Koh Samui, with an area of 26 mus. 70 villas were built inside the W resort hotel, with an area of 160 square meters for each one. They are also equipped with uniquely designed living rooms, bars and spas. There are 17 private properties inside the resort, with an area of 700-1300 square meters for each one. The total area is 1000 to 2500 square meters. Seascapes can be seen from afar inside the villas. These private properties are located at a 900-hectare secluded seaside area. Through the cooperation with MAPS Design Studio, the resort is equipped with the attractions designed by P49 Deesi—gn, a well-known Thai-based interior design company. And customers can immerse themselves in such a delicate and beautiful atmosphere of holiday life and enjoy leisure, relaxation, recovery and rebirth that cannot be found in elsewhere. We would like to invite you to our kitchen to taste the delicious food from all over the world or to the Namu restaurant to taste our creative Japanese cuisine, while enjoying the most spectacular view of the hotel——"the verge of fairyland." While you are taking

a break at the W beach, our Beach Ambassador will offer you all kinds of warm services. You can also engage in various aquatic sports. Head toward the place of leisure AWAY Spa Center, where you will enjoy all kinds of natural therapies soothing the body and mind to restore your sensory balance. At the same time, you can also try the only Mexican Temascal's special therapy called Thaimazcal. You can also have exercises by using the first-class equipment at the SWEAT Fitness Center, engaging in sports such as tennis, Muay Thai, yoga and aerobics that can energize yourselves. Or we can go for a swim in the WET Swimming Pool, enjoying cool, comfortable and luxurious experience.

小盘谷精品酒店

XIAOPANGU BOUTIQUE HOTEL

设计　黄靖

设计单位：北京华清安地建筑设计事务所有限公司
　　　　　黄靖联合建筑设计（北京）有限公司
参与设计：金楠、陈卫新、由学来、陈挥、宋溪
项目地点：扬州
结构/机电设计单位：北京中元工程设计顾问有限公司
建筑面积：3620 m^2（原有建筑 1850 m^2、新建建筑 1670 m^2）

小盘谷的名气虽大，然而喜爱园林的人却很难寻其芳踪，这座私家宅院与园林在近几十年中可以说是历经坎坷，深藏在扬州古城的曲折小巷中。

小盘谷位于扬州古城东南，南河下历史文化保护区，临近何园。其范围南起大树巷，北至丁家湾，东西均与民居相邻，占地面积约为3650 m²。小盘谷总体分为三部分：西部为三进平房宅院；中部为厅堂，分为过厅、主厅和后楼；厅堂两侧用火巷分隔，东侧火巷以东即园林部分。园林又分东西，走进上书“小盘谷”的园门，即为西园。园中有湖山颓石，旧名为“九狮图山”，因其山石外形如群狮探鱼而得名。山下有洞，洞出西口，有池水一泓，池上架石梁三折。池西水阁凉厅，三面临水，山洞北口，临水设石阶，石上嵌“水流云在”。东西花园以走廊和花墙分隔，墙南一桃门，上题“丛翠”，进桃门为东园，园南有凉厅三间。整个园林是以小见大之手法中最杰出者。住宅北侧，也就是文物保护的控制范围，曾经房管局修建了一座四层小楼作为招待所，高度、体量、风貌都对传统园林建筑有极大的影响，本次设计中将对其拆除重建。

如何将小盘谷的魅力原真展现，同时又使得传统建筑达到最大限度的使用？我们经过反复研究，以“扬州一日，梦已千年”时空穿越的主题，营造出现代园林会所的保护——改造——经营的模式。在厅堂中展览、在园林中游赏休憩、在宅院中居住，另外在西北侧增加一座两层的小建筑作为会所酒店的配套功能。

由于小盘谷是全国重点文物保护单位，我们在文物的范围内严格按照文物

保护法的相关要求，在扬州市政府、文物局、规划局、房管局的大力支持下，将园林与宅院根据传统记载进行修复，并在2009年9月完成，重现了“丛翠问茶”、“曲尺观山”、“桂花望月”、“桐韵修心”的精致园林景致（注：这四处分别是丛翠馆、曲尺厅、桂花楼、桐韵山房四组园林建筑）。在中轴线的厅堂部分，从大门、序厅、中堂、藏书楼，展现了周馥的生平事迹，大堂的匾额为当年慈禧太后御赐“风清南服”。西侧住宅部分和北侧改建建筑一起成为会所，将传统院落住宅和现代生活方式结合，创造出“一眼千年”的时空交错感觉。新建部分建筑面积约为1670 m²，地上2层1260 m²，地下局部1层410 m²：首层功能为会所酒店的入口、大堂及休息区、画廊、餐厅（仅为1个包间）；二层为5间客房（4个标准大床房，1间套房），及水池庭院、水榭、小戏台；地下室为设备用房、厨房等。三座保留下来的宅院由于房间进深较小，庭院较大，都在功能上作了调整：第一进院“听竹”为spa，第二进院“观鱼”。第三进院“问松”为两处不同风格的套房。

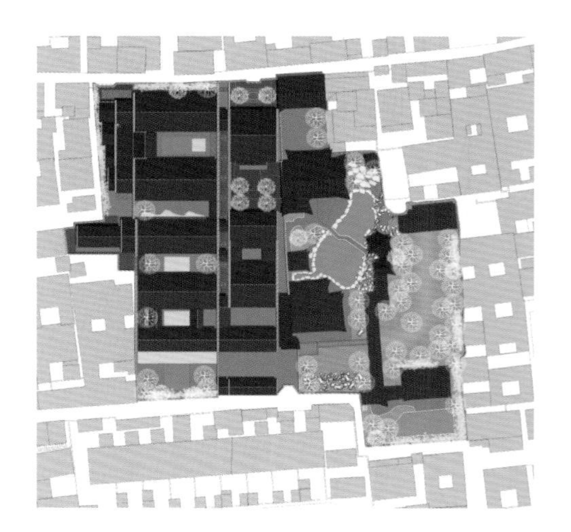

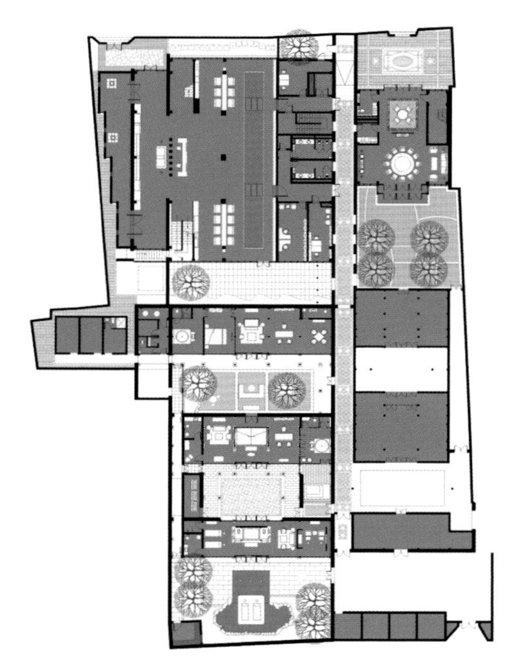

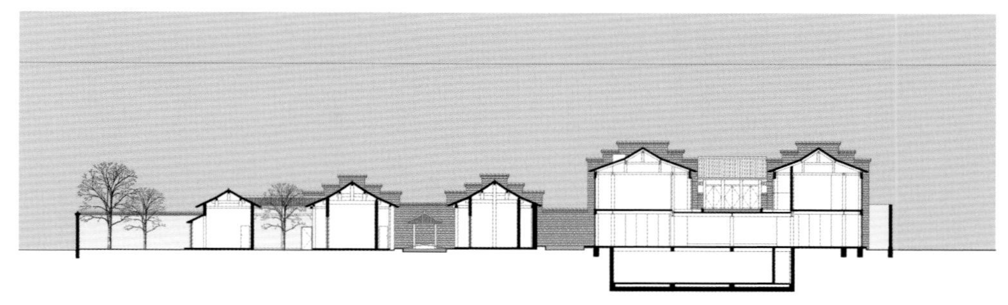

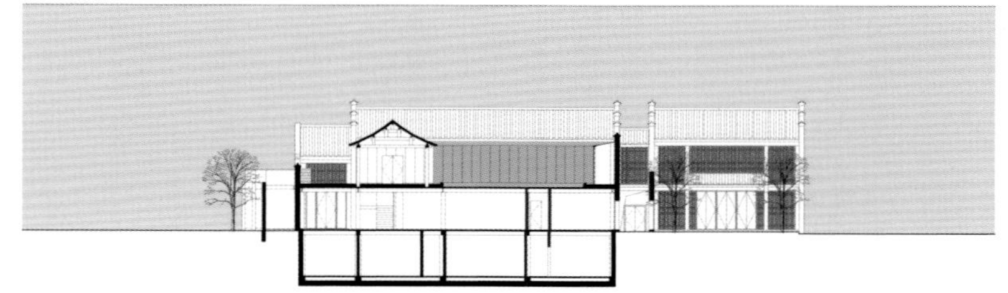

Xiao Pan Gu is famous, but those who are interested in gardening have difficulty to look for it. This private residence has undergone many twists and turns for decades and it lies deep inside the zigzag alleys in the ancient city of Yangzhou.

Xiao Pan Gu is located in the southeast of Yangzhou city, a historical and cultural reserve in Nan He Xia near the He Garden. The Da Shu alley is in its south, Zhang Jia Wan in its north. Its east and west are adjacent to folk houses. Now Xiao Pan Gu covers an area of 3650 square meters and it divided into three parts. There are three bungalow houses in the west. The center is the hall, and it divided into the hallway, the main hall, and the rear hall. There is a gap at the two sides of the hall. The garden area is in the east of the eastside gap. The garden area is also divided into the east and the west. When heading to the gate of the garden, there is the west part of the garden. There is a mountain-like rock in the garden. It was called "Nine Lions on Mountain" because of its shape where nine lions watch the fish in the lake. There is a hole at the foot of the mountain. At the west side of the hole, there is a lake. At the north side of the hole, there are stone steps near the water. A Chinese word "shui liu yun zai" is embedded on the stone. The east garden and the west one are separated by corridors and flowery walls. There is a gate with peach flowers and it's titled "cong cui". Getting into the gate, there is the east garden. In the south of it, there are three halls. The entire garden is the best piece of work by using small things to reflect big things. At the north side of Xiao Pan Gu, which is the relic protection site, was once a four-story reception office of China's housing administra-

tion. Its height, measurements, styles and features had a tremendous impact on traditional garden architecture. We will completely renovate it in this design..

We have had many discussions on how to show Xiao Pan Gu's real attractiveness and in the meantime maximize the use of traditional structures. Based on the time travel idea "One day in Yangzhou like a dream that lasts 1000 years", we will design a modern garden club and protection, renovation and operating models. Tourist can walk around in the hall, go sightseeing and take a rest in the garden and live in the residence. In addition, there is a tiny two-story structure at the northwest side of the residence, providing the hotel's complementary services.

As the major national cultural relic protection site, we strictly followed related requirements of the Culture Relics Protection Law based on the scope of the relics. Supported by Yangzhou municipal government, the State Administration of Cultural Heritage, the Urban Planning Administration and the Housing Administration, we restored the garden residence based on historical records and completed in September, 2009. We reproduced Cong Cui house, Qu Chi hall, Gui Hua building and Tong Yun mountain house. In the hall, where the central axis line lies in, from the gate, hallway, central hall to the library, Zhou Fu's major achievements are shown to the tourists. In the hall, there is the stele "feng qing nan fu" bestowed by Empress Dowager Cixi.

The hotel is consisted of the west side of residence and the renovated structures in the north. It combined the traditional court-style residence with modern lifestyles and formed a fast-changing and time-crossing feeling. New structures cover an area of 1670 square meters. The area of the two stories above the ground is 1260 square meters. The basement covers an area of 410 square meters. The entrance, the hall, the lounge area, the gallery and the restaurant are on the first floor. There are 5 guest rooms (4 standard rooms and a suite), a courtyard with a swimming pool, a waterside pavilion and a stage on the second floor. The basement is equipped with a maintenance room, a kitchen etc. We made adjustments on the functions of the three residences because their rooms are small and their courtyards are large. The front area "ting zhu" is the spa, the central area "guan yu" and two different suites are in the back part of the hotel.

扬州三间院二期四水堂精品酒店

A SECOND OPERATION OF SISHUITANG BOUTIQUE HOTEL AT SANJIAN COURTYARD, YANGZHOU

设计　黄靖

项目地点：江苏扬州市
设计单位：黄靖联合建筑设计（北京）有限公司
业　　主：扬州泰达发展建设有限公司
参与设计：金楠、蒋晓春
建筑面积：2750 m^2
竣工时间：2010

三间院的得名始自南京大学的张雷老师，他在一期工程设计时，以扬州园林中的水、竹、石山中元素作为三组院落的主题，并以南京乡间传统农居建造方式完成了三组建筑，成为扬州独具一格的餐饮会所。一期餐饮建筑的成功使得二期酒店的设计要求更高，不仅要很好的延续三间院的水、竹、石元素意境，同时要与一期风格有明显的差异。由于项目时间很紧迫，设计时间不足1个月，而且要求低造价、短工期；好在业主方给与我们创作想象的空间，并全力支持与配合。于是我们在2009年12月开始设计，2010年1月完成施工图和室内设计，春节前开工建设，4月16日建成使用，也算是扬州新城的神奇速度吧。

三间院位于扬州市广陵新城的东侧，廖家沟西岸边的绿化景观带中，环境优美，完全与喧嚣城市隔离。二期四水堂酒店在一期餐饮建筑南侧，以一条小溪相隔。平面布局采用中国传统的九宫格，方正严谨，突出传统建筑的轴线、规制、私密感。守在四角的四组院落围绕正中间的大厅，并用廊桥将五组建筑相连，简单有效的布局使得后期建造过程十分的顺利。

三间院二期建筑，依然延续了对扬州园林建筑的提炼与升华，继续在“水、竹、石”上做文章。四水堂酒店坐落在一片水面之上，白墙黛瓦映于水中；而连接建筑之间的廊桥和水中央的大厅均用竹子包裹，室内的装饰主材也用竹饰或复合竹材；整个建筑四个立面极少开窗，均向内院或侧院开窗采光，取义石材的方正简洁。

四水堂酒店从开工建设到投入使用只用了80天的时间，有些设计的深化与修改只能和施工同步进行，特别是室内设计功能布局。面向夕阳的酒店入口是气派的竹廊，透过竹柱的开敞空间可以看到水面中央的大厅；左侧西北院落为对外服务区，包括多功能厅、会议室、贵宾休息室和服务间，有服务通道可以便捷地和一期厨房相连；右手西南院落是

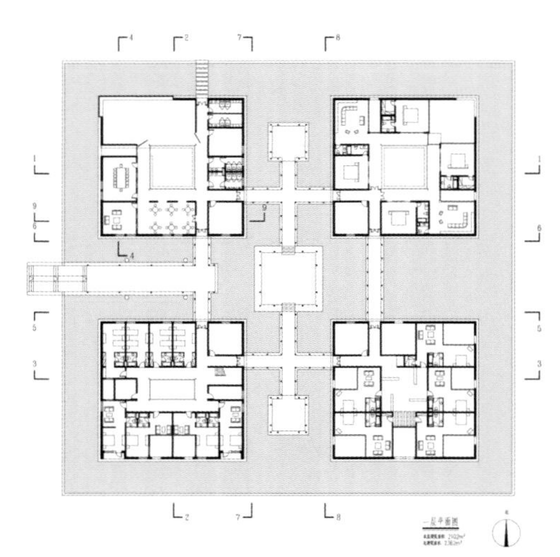

标准客房区，局部两层共有12间标准间；东北院落是两套豪华客房，各自的"L"形房间组合成方正的院落；东南院落为5间特色院落套房，每套有专属的庭院可透过院墙欣赏外面的水乡景观。四组院落围合的水中央的竹厅是酒店的大堂酒吧，客人可以倚靠在水边谈天说地。南北两侧的水中还设有两处竹亭，可提供会谈或茶饮。

四水堂酒店在定位准确的前提下突出了"快"，但由于对市场研究的充分，以及三间院一期的影响，使得酒店建成之后获得良好的口碑；而且施工迅速、材料简单、格调高雅、配套完善，使得业主方经营得到了预期的回报。

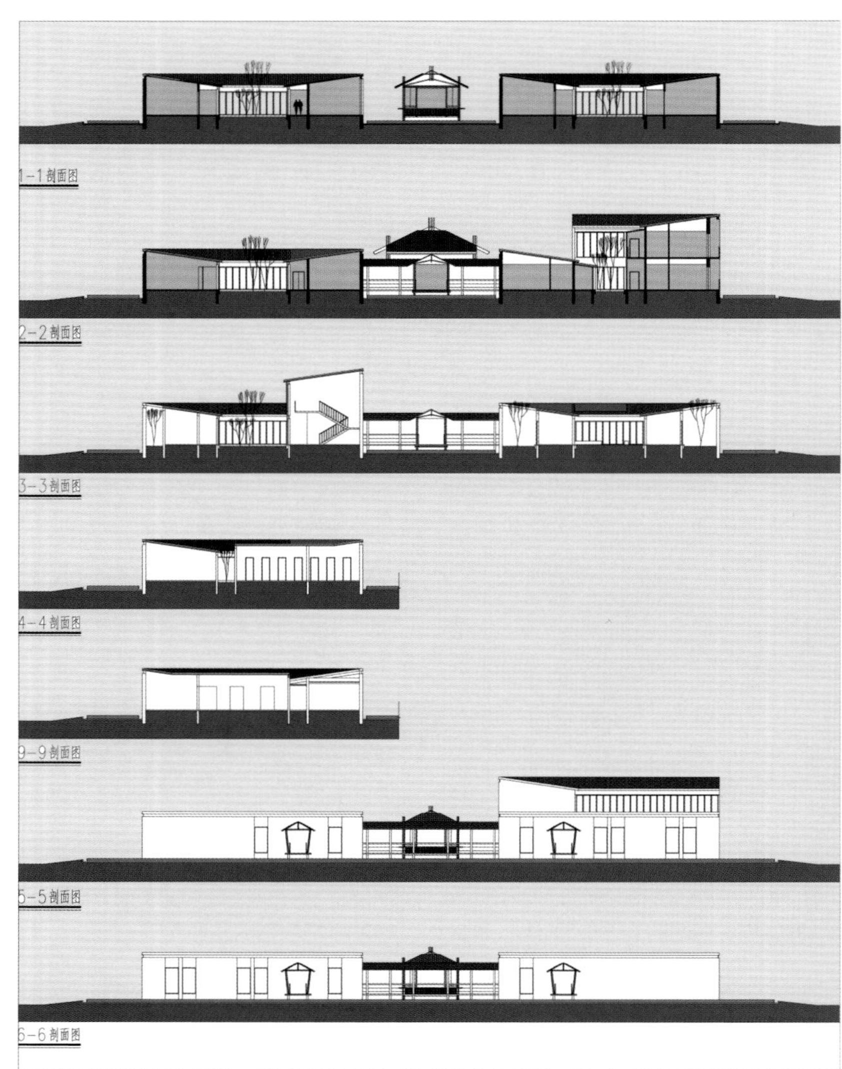

San Jian Yuan is named by Zhang Lei, a teacher at the Nan Jing University. In his first-phase design, water, bamboo and stone in the Yangzhou garden are the main elements. The residence, with a rural style with Nan Jing's tradition, is a unique restaurant in Yangzhou. The success of the first phase set a higher bar for the hotel in the second-phase design. It needed to carry forward the elements of San Si Residence, and also be more creative over the first-phase design. With less then a month's time, the project was very pressing. Lower costs are required and it needed to be completed in a short span of time. Thanks to the owner's support and cooperation, we had an opportunity to extend our creativity and imagination. Started in December, 2009, we finished the construction drawing and interior design in January, 2010. The construction had already been underway before 2011's Spring Festival. It opened for business on April 16th, 2011.

San Jian Yuan is located in the east side of the Guang Ling Xin Cheng in Yangzhou city and in the green

area in the west coast of Lu Jia River. It' s beautiful and is completely isolated from the bustle of the city. Si Shui Tang Hotel is in the south of the restaurant, separated by a stream. The floor plan is carefully designed as a square with Chinese tradition, highlighting the axes, specifications and privacy of traditional structures. The central hall is surrounded by the four reeves' shad statues in the corners. The five groups of building are connected by bridges. Such a simple and effective layout benefited the construction in the final phase.

Si Shui Tang Hotel still carries forward the refined excellence of Yangzhou garden. Water, bamboo and stone are still the main elements in the hotel' s design. Si Shui Tang Hotel lies on the surface of the water, which reflects the shadows of its white walls and bricks. The bridge between the two structures and the central hall are bamboo-decorated. The interior is also decorated with bamboo. The windows in the exterior are usually closed. The ones in the interior and the two sides are open to let the sunshine in. The stone material is foursquare and simple.

It took 80 days to build the hotel and start operations. Some design and improvement could only be finished during the construction, in particular the interior layout. The hotel entrance facing towards the setting sun is the magnificent bamboo colonnade. The central hall on the surface of the water can be seen through the gaps between the bamboo poles. The northwest part in the left is the service area, including the multi-functional hall, conference room, VIP lounge room and customer service area. It is connected with the restaurant. The guest rooms are in the southwest of the hotel. The two-story building is equipped with 12 standard rooms. There are two luxurious rooms in the northeast. A L-shape arrangement of rooms form a square courtyard. There are 5 suites in the southeast, with a courtyard in each of them. The waterside landscape can be seen through the walls. The bar is in the central hall surrounded by these four big courtyards. Customers can sit by water and talk to each other. There are two pavilions in the south and the north where people can have a talk and enjoy tea.

Si Shui Tang Hotel and Xiao Pan Gu are carefully designed structures. They highlighted "jue" based on their specific goals. The research we have conducted into the market and the impact from the San Jian Yuan Restaurant brought huge benefits to the hotel after it opened for business. And the owner gained expected results due to the hotel' s expeditious construction, simple construction material, graceful style and complete complementary services.

亚布力 Club Med 冰雪度假村

CLUB MED YABULI

设计　Marcelo Joulia

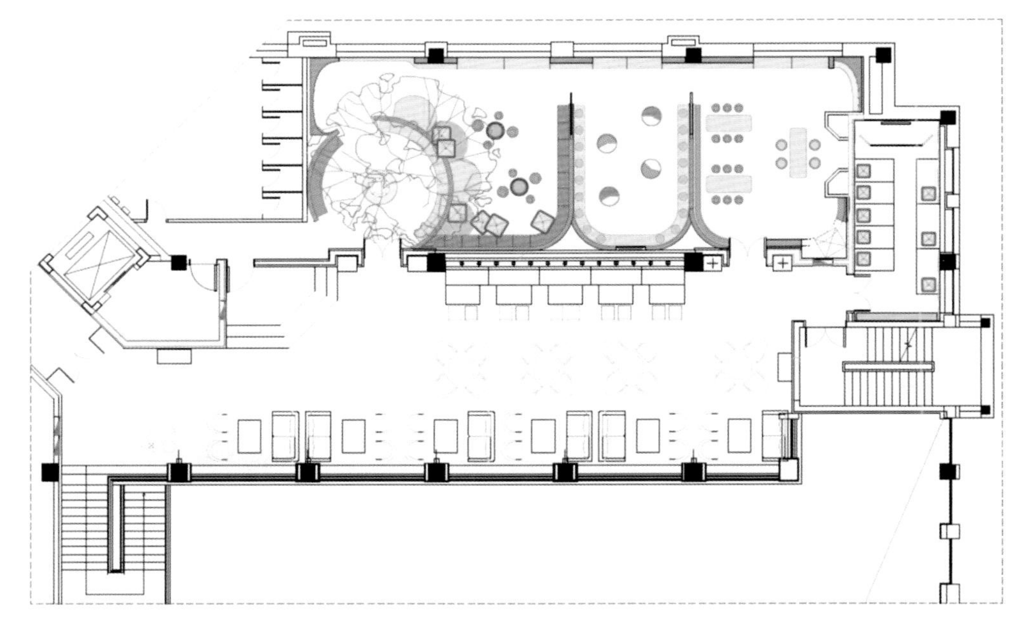

作为国际度假集团Club Med在中国的首个度假村，由纳索建筑设计事务所设计的Club Med亚布力冰雪度假村提供包括住宿、餐饮、交通、娱乐等"一价全包"的服务，酒店内热火朝天的娱乐活动及室外冰天雪地的景色让客人体验何谓冰火两重天。

在这个位于黑龙江省东北部的地带，雪季几乎占了全年的一半，冬天极度寒冷。才下午四点多，亚布力已经入夜，我们根据轮胎碾过地面的声音判断积雪的厚度，对即将到来的雪上狂欢充满期待。当夜色已经完全黑透的时候，我们终于到达了此行的目的地——Club Med亚布力冰雪度假村。

在黑夜之中，度假村酒店的楼群犹如一座冰雪城堡，而前来迎接的酒店员工却是人人都红光满面，还没有来得及体会冰寒，我们率先领略了亚布力的热情。作为Club Med的首座中国度假村，亚布力冰雪度假村也和全球其他Club Med度假村一样，提供包括住宿、餐饮、交通、娱乐等"一价全包"的服务。Club

设计单位：Naco Architectures（纳索建筑设计事务所）
项目地点：黑龙江亚布力
建筑面积：30000 m^2
摄 影 师：Jonathan Browning

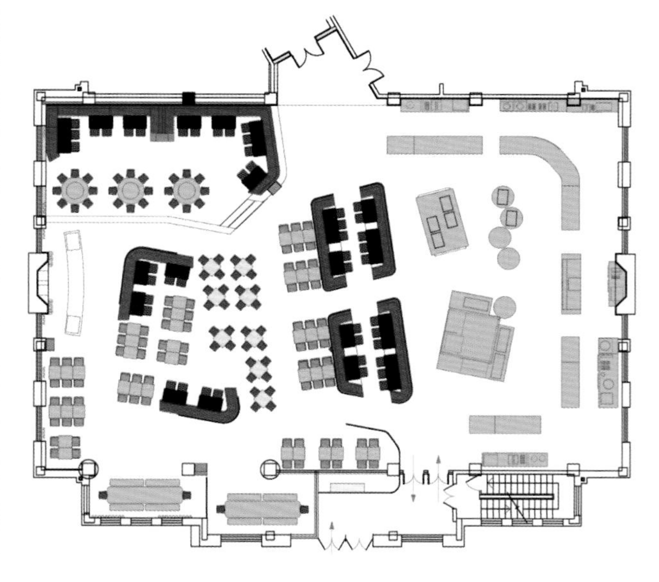

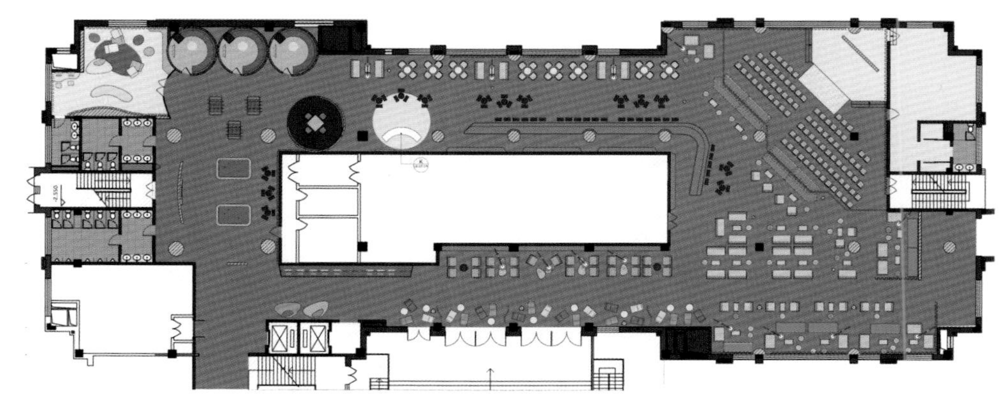

Med的工作人员有个特殊的称呼叫作G.O(Gentle Organizer),由来自世界各地的青年男女组成,他们既为客人服务,又带动和组织客人参与各项活动。一名会讲韩语的帅小伙成为了我在酒店认识的第一位G.O,他把一条热毛巾、一杯姜茶以及一份详细的住宿说明书递到我的手里。

来到Club Med,你必须先阅读这张说明书,不然很可能会错过许多精彩的活动,比如舞蹈课程、瑜伽训练、有氧操、乒乓球、桌球、麻将、蹦床以及各种为儿童设置的娱乐项目,甚至还有你根本没想过会出现在酒店的空中飞人马戏课程。当然,这些全部是免费的。所以,在酒店中经常可以看到大家在酣畅淋漓地做运动的场面。除此之外,酒店还为喜欢夜生活的人提供了一个热闹的场所——森林酒吧。酒吧的整体设计以亚布力所在的五花山为灵感,家具采用山上的五种色调,其间错落分布着树皮壁纸覆盖的立柱,专门定制的地毯图案来自五花山地表图,而吧台则是对传统中药橱柜的重新演绎。夜晚的森林吧是热火朝天的场面,客人不但可以免费品尝各式调酒,还能唱卡拉OK,或者在G.O的带动下一起跳舞。当然,对很多客人来讲,来到亚布力就一定要和寒冷交战,那么露天按摩浴池就是个直面寒冷的地方,你将在冰雪的包围中看热气袅袅上升,体验何谓真正的冰火两重天。

虽然丰富多彩的娱乐活动让客人根本不想睡觉,但客房仍旧值得一说。酒店的客房十分宽敞,即便是标准间也毫不吝啬对空间的使用,棕色厚绒地毯搭配褐色木制家具,让空间充满暖意,壁纸上的树皮纹路及窗帘上的枝叶图案则呼应了树林的主题。踏进洗浴间内,一股热气顿时从脚底袭来,原来这里安装了地暖系统,类似设施还有公共区域的电子壁炉,让人对酒店的供暖印象深刻,单从温度上来讲,你绝对感受不到室外正值北国的冬天。

翌日清晨,当你拉开客房的窗帘,一定会被眼前的画面所震撼。此时刺痛你眼睛的不仅仅是强烈的阳光,还有一片白茫茫的冰雪世界。透过巨大的玻璃窗,能看到蜿蜒的雪道和往来不息的缆车,顺着雪道及缆车的方向望去山顶,技高人胆大的滑雪者们正尖叫着俯冲而下,身后溅起一路雪粉,惊险如武侠片中的打斗场景。此时,我也迫不及待要投入到这个度假村内最"热"的活动中,去体验那彻头彻尾的寒冷了。

As the first holiday village of the international holiday group Club Med in China, the Club Med Yabuli snow resort, designed by Nassau architectural design firm, provides a "one-price-all" services including accomodation, catering, transportation, entertainment and so on. The bustling entertainment and outdoor ice and snow scenery have let people experience the icy and hot situation in personnel.

At northeast part of Heilongjiang Province, the snow season has covered almost half of the year. Its winter is cold extremly. At four o'clock in the afternoon, Yabuli has enter its night time. We could judge the thickness of the snow in accordance with the noise from the tire running over the ground. We are crazying about the coming snow. When it is dark night, we arrived at the destation--Club Med Yabuli snow resort.

In the dark, the resort hotel seems like a snow castle. The waiters here are all warmly. We have first enjoy their warmly welcome before experiencing the cold. As the first resort of Club Med in China, Yabuli snow resort is same with other countries club Med, providing a "one-price-all" services including accomodation, catering, transportation, entertainment and so on. All Club Med staff have a special name called G.O (Gentle Organizer), which is constituted with the young boys and girls coming from the whole world. They are customers and lead customers to participate in various activities. A young guy that can speak Korean is the first G.O i know in this hotel. He handled a hot towel, a cup of ginger tea as well as a detailed accommodation instructions into my hand.

Entering the Club Med, you shall read this instruction first, or you will miss a lot of wonderful activities, such as dance course, yoga trainning, aerobics, table tennis, billiards, mahjong, trampoline, and a variety of settings for children entertainment. There is even trapeze circus course which you have never thought about. Of course, all these courses are free of charge. Therefore, we could find the sweat people taking their sports. In addition, this hotel also provide a a lively place--Forest Bar for nightlife. The bar's overall design is taking Wuhua Mountain located in Yabuli as the spirit. The furniture is adopted five colors of the mountain. Among them, there are set pillars covered with tree-skin-pattern wallpaper. The special made carpet patterns are from the Wuhua Mountain surface diagram, while the bar table is re-interpretation of traditional Chinese medicine cabinet. At night, the Forest Bar is a bustlign scene. Guests can taste all kinds of wines here for free, as well as play Karaoke, or dance with G.O. Of course, for many guests, they have to have a battle with cold in Yabuli. Then, the outdoor Jacuzzi is the best place to face the cold directly. You will watch the hot steam raising to the sky in the snow place, let people experience the icy and hot situation in personnel.

Though the colorful entertainment letting people have no sleepy sense, the room is need to introduce also. The room of this hotel is spacious.Even for the standard room, its space is not less any more. The brown thick velvet coupled with the brown wooden furniture, has made the space full of warm

sense. The tree-skin-pattern wallpaper and the leaves pattern on curtains are echoing the theme of forest. Entering the bathroom, a stream of heat wave suddenly hit from the soles of the feet. Then, we find that there is installed a floor heating system, as well as the electronic fireplace in public areas. People will have a deep impress to this hotel. Just from the temperature point, you will never feel that the outdoor is in deep winter.

At the next morning, when you open the curtains, you must be catched by the scene. The shining sunlight and the icy and snowy world are sting your eyes. Through the huge glass window, you could see the winding snow road and coming&going cable car. Along with the snow road and cable car direction, you could find the skilled and daring slipping down. The snowpower is stirred behind, thrilling as the fight scenes of martial film. At this time, i can not wait to put in yo this most "hot" activity to enjoy the complete cold.

天津海河英迪格酒店

TIANJIN HAIHE INDIGO HOTEL

设计　Patrick Waring、Susan Heng

设计单位：Silverfox Studios
项目地点：天津

天津海河英迪格酒店是天津独一无二的时尚精品酒店，亦是中国唯一在市中心提供别墅式住宿的酒店。酒店地理位置优越，从天津乘高铁列车抵达中国首都北京仅需27分钟。秉承英迪格品牌的一贯传统，天津海河英迪格酒店为宾客提供令人耳目一新的本土体验，却又不失独特的现代气息。英迪格旗下的每座物业均为独立设计，不仅反映了这座城市的风貌，更融入所在街区的文化、特色及历史。天津曾经是中国最重要的通商口岸，在19世纪末及20世纪初，有九个国家在天津设立了自治的租界。这座城市悠久的历史风情以及充满活力的现代气息，均为英迪格酒店提供了丰富的灵感源泉。

英迪格酒店以精品体验及时尚生活作为品牌标志，并融合地方特色。天津海河英迪格酒店座落于1899至1917年的前德国租界之内，根据要求，设计需要将这历史地点和独特位置，与天津的现代风貌互相配合。由于该项目共有11幢别墅，我们可于其中注入不同理念，并保留德国租界特色。

主要的公共设施如接待处和餐厅均位于主别墅内。主别墅座落道路交汇处，是酒店显而易见的入口。其前身为德国领事馆，与1899年的天津市拥有深厚的历史联系。我们保留了原有建筑细节，并添上现代元素以构成鲜明对比。室内的艺术品展示了旧城地图及著名建筑，当中部分至今仍然屹立不倒，而且就在酒店附近。

此外，主别墅的室内设计理念亦包

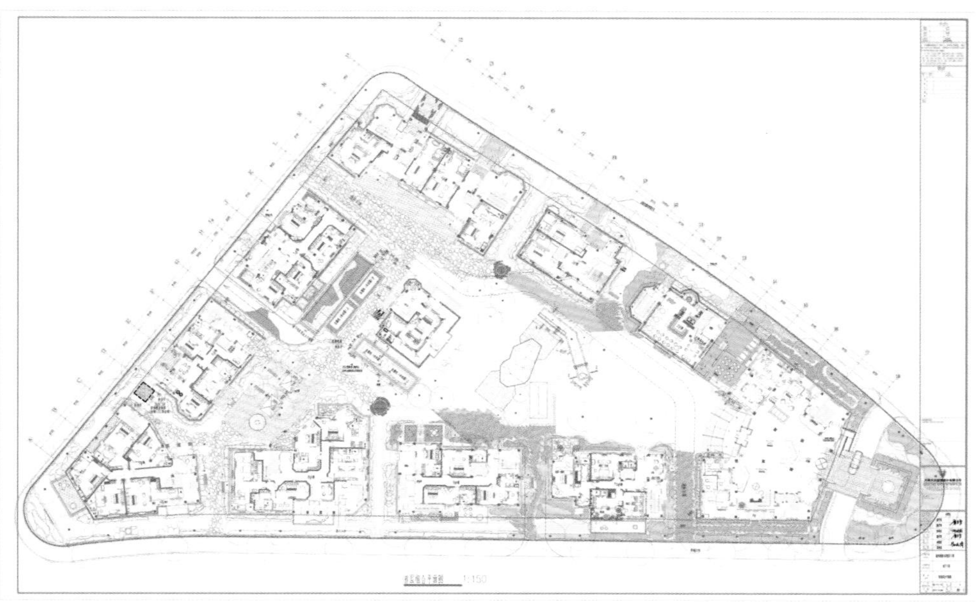

括1899至1917年间的邮票和信笺图像，以及昔日于老天津经营餐馆生意的著名家族。当年，会所乃重要的社交场地，而酒店现设有酒廊、附设壁炉的图书馆，以及惬意舒适的私人空间“Me Space”，让宾客尽享个人互联网服务及写意时光。

酒店以现代手法演绎欧式传统细节，作为贯穿整个项目的主要特色。镶板墙的造型不但带有古典气派，亦可配合当代设计。另外，我们亦于建筑木制品添上艺术元素，描绘出天津德国租界的历史故事。

特定的设计主题可见于整个发展项目上，例如：电影院主题客房灵感源自昔日于德国租界内设立第一间有声电影院的俄罗斯商人；而德国啤酒屋暨餐厅正是这位重要人物的故居。

天津河岸的商贸活动，是天津市设立租界及蓬勃发展的原因，我们将此带入设计当中，并于其他地方以天津的今昔对比作为设计理念。

每幢建筑的设计风格迥异，每层空间大少不一，部分房间设有烟囱，另一些则备有窗台或露台，这一切令酒店的每间客房配置与面积皆有所不同，堪称绝无仅有。

BEER
WINE

Tianjin Haihe Indigo hotel is the unique one of fashion and boutique, and also the only hotel which is located in downtown and offer villa type accommodation in China. With the superior geographical position, it only takes 27 minutes to reach in the Chinese capital Beijing from Tianjin by high-speed rail. In accordance with the consistent tradition of Indigo, Tianjin Haihe Indigo hotel offers the refreshing mainland experiences for customers without losing the unique modern flavor. Every real estate of Indigo is designed on its own, which not only reflects the style and features of the city but also further integrates the culture, features and history of its own block. Once the most important treaty port in China, there are 9 countries establishing self-governing concessions in Tianjin between the end of 19th and the early of 20th century. Both of the long historical amorous feelings in this city but also the modern flavor which is full of energy offers abundant inspirations source for the Indigo hotel.

Indigo hotel takes the boutique experiences and fashion life as the brand logo as well as blends the local color. Located in the former German concession which was once a concession from 1899 to 1917, on request that the design of Tianjin Haihe Indigo hotel should coordinate the his-

torical place and unique position with the modern style and features in Tianjin. Due to its 11 villas, we can inject different ideas into it and maintain the features of German concession.

The main communal facilities like reception and restaurant are both located in the main villa. Located at the intersection of roads, the main villa is the obvious entrance to the hotel. Formerly known as the German consulate, the main villa has deep historical relations to Tianjin which is in 1899. We maintain the original construction details and add modern elements to contrast with the original. The interior work of art shows the map of the old city and famous buildings, part of them still stand erect near the hotel.
In addition, the interior design philosophy of the main villa includes the images of stamps and letter paper from 1899 to 1917 and the fame families who ran restaurants in old Tianjin in the past.

In those days the club is the important social contact area, while the hotel now is equipped with bar, a fireplace is attached to the library, satisfied and comfortable personal space "Me Space", all of them enable the guests to enjoy personal internet service and comfortable time.

The hotel deduces the European style traditional details with modern techniques to be the predominant feature which run through the whole project. The modeling of bordering with siding walls not only carries with the classical style but also coordinates the contemporary design. In addition, we also added art elements to wooden products of buildings to trace out the historical story of the German concession in Tianjin.

The special design themes can be visible on the whole developing projects, for example, the inspiration of the theme guest rooms originates from the Russian merchants who set up a first voiced cinema in German concession in the past; while the German beer house, a surname of restaurant is just the former house of this important figure.

The riparian business activities in Tianjin are the reasons for establishing concessions and flourish, which will be brought into design. Comparing the present with the past serves as the design philosophy as other places.

Distinctions in the design style of every building, different space of every floor, partial rooms have chimneys. Some of the others are equipped with windowsill and terrace. All of them make every room of the hotel being different from both of the configuration and area, called one and the only one.